Industrial Research for Future Competitiveness

Springer
*Berlin
Heidelberg
New York
Barcelona
Budapest
Hong Kong
London
Milan
Paris
Santa Clara
Singapore
Tokyo*

Klaus Brockhoff

Industrial Research for Future Competitiveness

With 20 Figures
and 13 Tables

HD
30.4
.B773
1997
West

Springer

Prof. Dr. Klaus Brockhoff
Universität Kiel
Institut für Betriebswirtschaftliche
Innovationsforschung
Olshausenstr. 40
D-24098 Kiel

ISBN 3-540-62842-8 Springer-Verlag Berlin Heidelberg New York

Die Deutsche Bibliothek – CIP-Einheitsaufnahme
Brockhoff, Klaus: Industrial research for future competitiveness: with 13 tables / Klaus Brockhoff. – Berlin; Heidelberg; New York; Barcelona; Budapest; Hong Kong; London; Milan; Paris; Santa Clara; Singapore; Tokyo: Springer, 1997
ISBN 3-540-62842-8 Gb.

This work is subject to copyright. All rights are reserved, whether the whole or part of the material is concerned, specifically the rights of translation, reprinting, reuse of illustrations, recitation, broadcasting, reproduction on microfilm or in any other way, and storage in data banks. Duplication of this publication or parts thereof is permitted only under the provisions of the German Copyright Law of September 9, 1965, in its current version, and permission for use must always be obtained from Springer-Verlag. Violations are liable for prosecution under the German Copyright Law.

© Springer-Verlag Berlin · Heidelberg 1997
Printed in Germany

The use of general descriptive names, registered names, trademarks, etc. in this publication does not imply, even in the absence of a specific statement, that such names are exempt from the relevant protective laws and regulations and therefore free for general use.

Hardcover-Design: Erich Kirchner, Heidelberg

SPIN 10575845   42/2202-5 4 3 2 1 0 – Printed on acid-free paper

Foreword

Business leaders talk about it and the statistics show it: the 1990s are not years of booming research expenditures in either the European or U.S. corporate environments. Even in Japan, industrial research seems very recently to have bowed to short term economic troubles. This is astonishing for at least two reasons.

First, discussions regarding the benefits and the necessity of innovations have not decreased in intensity. However, few participants in these discussions recognize an operational relationship between research and innovation, if only because the terms tend to be used almost arbitrarily in public debates. Secondly, an increase in total Japanese involvement in research is clearly visible. Both of these observations could be without value. The Japanese decision makers could be wrong, and there could be no coupling between research and innovations. On the contrary, the Japanese decision makers could be on the right track and the 'Western' business leaders could be wrong, if such coupling does indeed occur. At least the necessary conditions for the success of the coupling should be outlined. Besides, it could be asked whether research in companies contributes to build other potentials than those leading directly to innovations. If this could be made plausible, it would represent yet another reason for supporting research. These are some of the issues addressed in the following text. Obviously, they are of great importance in understanding the competitiveness of firms, and perhaps even regions. These were compelling reasons to look into this problem area.

Looking back at the study, I conclude that giving up research in a technology-depending business means gambling away future competitiveness. This could be further aggravated if government supported research institutions succumb to pressures to give up their basic research as well. In some countries, this is exactly what happens.

The present writer is convinced that enough evidence is available to justify company expenditure for research. It is a discomforting observation that recent waves of management fads do not quite favor long-term, risky research investments. To what degree firstly the installation of managers with short-term performance evaluation and secondly the advent of consultants with even shorter term interests in demonstrating the success of their work can explain the recent downrating of research in "western" corporations is an open question. We do not set out to answer it. Perhaps, the observed actions of managers and consultants were "justified" by a lack of consistent research management. In fact, managing research is rather difficult, and in this study we hope to contribute a few ideas to improve this management.

Although the topics we want to discuss here would appear to be of recurring interest to business people, relatively little has been written about them in the business literature. Within the extant writings we discover not only increasing understanding and refinement, but also revisions of earlier views. This was yet another reason to deal with research in the company environment. We hope that we have integrated the major findings, and present them in a way that is helpful for practitioners and that spurs interest of business researchers in further expansion of the knowledge level.

This publication has benefited from discussions with members of the Company of the Future-Project and with participants of the R&D Management Conference 1995 on "Knowledge, Technology and Innovative Organizations", Pisa/Italy. I wish to thank the discussants for their comments. The research was supported in part by the European Commission, DG III, ESPRIT Program, Technology for Business Processes, project 9803. Among other things, this support enabled Dipl.-Kfm. Justus Bardenhewer to interview 19 European and 24 Japanese technology managers. Some of his results are integrated into this text, based on his more elaborate texts that will appear elsewhere. The technology managers represented in the EC project were patient partners in discussions and great supporters in setting up contacts both within their companies and their countries. We benefited particularly from the extensive experience of the research directors of major electronics companies, namely Drs. A. Airaghi of Finmeccanica, H.G. Danielmeyer of Siemens, T. Nakahara of Sumitomo, and Y. Takeda of Hitachi. Dr. Olaf Eggers was my doctoral student, who cleared some of the ground in his dissertation. Dr. H. Ernst performed patent analyses that will result in separate publications, and that are only summarized in this text. Doctoral research by A. Bender, V. Lange, U. Pieper, H. Sattler, G. Schewe, J. Vitt, A. von Boehmer and K. Waschkow is quoted in our text. It is a further proof of the innovative and interesting ideas that originated from the

„Graduiertenkolleg Betriebswirtschaftslehre für Technologie und Innovation", a Ph.D.-program sponsored by the Deutsche Forschungsgemeinschaft and the State of Schleswig-Holstein. I am extremely grateful for this sponsorship since it helped to create a truly extraordinary research focus that today qualifies as a core competence of our Institute. Ms. C. Pöhler and Mr. R. Fraundorfer were the graduate students who helped with the identification of interviewees and data collection. Ms. C. Huch corrected the text, and Ms. D. Jensen was more than helpful in putting all the bits and pieces into a readable manuscript. My sincere gratitude goes to all of them. I do not know, whether all of those mentioned share my views. But I do hope that they agree with the major conclusions of this study. As always a number of unclear passages and even failures will remain. The author alone assumes full responsibility for these shortcomings.

Institute of Research in Innovation Management
University of Kiel, Germany   Klaus K. Brockhoff
January 1997

Contents

1. The crucial question: to invest or not to invest in research? 1

2. Characteristics of research 13
2.1 Problems of measurement and definition 13
2.2 Appropriability and transferability of research results 19
2.3 Specific risks of research 26
2.4 Specific benefits from research 28

3. Reasons for research 31

4. Research as a source of potentials 41

5. Research potentials and project funding decisions 51

6. On property rights and project potentials 61

7. Research potentials and relative share of research 63
7.1 Mandatory research 64
7.2 Mandatory research and transfer cost 66
7.3 Supportive research 68
7.4 Tests of the basic relationships 68
7.5 Estimates of research elasticities 71
7.6 Limits to research expenditure 76

8. On sufficient conditions for research success 79

9. Primary research potentials as a necessary condition for research success 85
9.1 A taxonomic approach 85
9.2 Diagnosis of trouble 94
9.3 Suggestions for building potentials 99

9.4 How much attention for each function?...............................110

10. On locating research..113
10.1 Qualitative analysis...113
10.2 Modeling the location decision...119

11. Conclusions...123

Appendix

Basic research expenditures as a percentage of total research and development expenditures in major industrialized countries, 1971-1993 131

Nominal and real industrial research expenditure in three countries (Germany, Japan, U.S.) 132

The relationship between the share of industrial research expenditures and the interest rate in Germany for three industries, 1965-1991 136

Example of a Mission Statement 137

Literature 139

List of Figures

1. Basic industrial research as a share of the total industrial R&D expenditure in three idustrialized countries 14
2. Basic and applied industrial research as a share of industrial R&D expenditures in Japan and the U.S., 1981-1993 18
3. Share of direct industrial research funding and level of industry's influence on the choice of topics in publicly funded research institutions, such as universities 22
4. Functions of corporate research 33
5. A comparative overview of studies on functions of industrial research 40
6. A sketch of our concept 42
7. Distribution of sources of R&D funds 49
8. The system of company research potentials 50
9. Example of funding by projects and potentials 54
10. Project funding form 57
11. Interdependencies between Nokia Central Research (NCR), top management and the business units (BU's) 58
12. Finmeccanica's process of technology planning 60
13. Relative impacts of relative budget changes (Elasticities) 75
14. A sketch of the process for securing sufficient conditions for research success 84
15. Relevance and accessibility of external sources of research in Germany 103

16. A process view for improving the necessary conditions for research performance.................................... 109
17. Views of the budgeting process 127

List of Tables

1. Major basic research laboratories in the Japanese electronics industry 11
2. Functions of research in 26 German companies 35
3. Comparison of German and U.S. perceptions of research functions 36
4. Actual and ideal performance of research functions as seen by 17 research managers in Europe and 20 in Japan 37
5. The relevance of scientific fields to technology in the U.S. 39
6. Funding procedures in central R&D laboratories of 53 large German companies 53
7. The relationship between the share of industrial research expenditure and the interest rate in Germany, 1965-1991 70
8. A characterization of necessary conditions for research success with low relative technological position 88
9. A characterization of necessary conditions for research success with high relative technological position (1) 89
10. A characterization of necessary conditions for research success with high relative technological position (2) 90
11. A characterization of necessary conditions for research success with high relative technological position (3) 92
12. A characterization of necessary conditions for research success with high relative technological position (4) 93

1. The crucial question: to invest or not to invest in research?

Should firms allocate a portion of their funds to research, seeking new technological knowledge that may not find immediate application in products or processes? Recently, more managers have been answering 'no' to this crucial question, because they feel that
- results are not observable after relatively short planning horizons have been adopted
- results cannot be fully appropriated by the firm that spent to obtain them
- research proves to be risky
- public institutions engage in research that is of potential value to the firm anyway.

While these arguments appear to be strong, we feel that the answer cannot be found so easily.

First, enlightened business leaders are well aware of the necessity to take long term aspects into consideration. Namihei Odaira, the founder of the Hitachi, Ltd., and its research laboratories, wrote in 1930 in a calligraphy: "Though we cannot live one hundred years, we should be concerned about 1000 years". He thus advises management to adopt a long term perspective and to approach goals steadily. Managers cannot take the advice literally, as this would imply an almost unlimited time horizon for the planning of their activities. It would also imply to invest very large sums of research money for exploring the future. Diverging from Odaira's advice, it is economically rational to limit the planning horizon much more strictly. This implies that much smaller sums would be invested into research, if research should be performed at all.

It may be frustrating for research managers to observe that planning horizons are drastically shortened in response to more environmental turbulence. In fact, the Ifo-Institute, a leading German economic policy research institute, reported that only 2.6% of close to 1,500 respondents to a questionnaire sent to 10,000 German compa-

research and development planning (Ifo, 1990, p. 2). Even if research leads to future returns these are easily cut away by the employment of such short planning horizons.

Yet it is possible that planning under limited time horizons may still keep the longer term objectives in perspective, and thus retain a certain level of resistance to short term changes. An interesting example of this is provided by K. Matsushita in a speech delivered in 1932 to his employees. He sketched objectives for his firm and for the electrical industry for a time frame of 250 years, but then he proceeded to break down this period until he arrived at the eight years immediately following and the advice that he had to offer for this period. Here, a culture is established that buffers short term fluctuations and offers a long term perspective for research expenditures.

Second, resource allocation to any company activity makes sense only if it generates an income large enough to cover the risk involved in the investment expenditure. Results from research, unlike many development results, cannot easily be appropriated by the institution that supports the research work because these results can hardly be protected from their use by rivals. This is a central argument in economics that explains private underinvestment in R&D, and which is used to call for government subsidies (Nelson, 1959; Arrow 1962). It may sound ironic that the deficiency of any one firm to appropriate the results of its research and the possibility of a spillover of these results to other firms is considered a substantial growth factor in more modern economic growth theories (Grossmann, Helpman, 1991). However, we do not wish to assume an economic perspective. Rather, we are concerned exclusively with a company-centered approach.

Another reason for not fully appropriating research results to the sponsor of the research effort may be that market conditions do not allow a profit to be made from the research results. If a dominant product design rules the market or if a large installed product base creates network effects or similar externalities, even very attractive alternative research results may not lead to the market success of an individual supplier. The major reason to perform company initiated research under these conditions may be to enter entirely new markets that are not similarly restricted. However, this increases the risks involved in research, as entering new markets requires managerial and other skills that are apparently very scarce. This is demonstrated time and again by failed attempts at diversification, even in cases where the technological base is already available. Thus, only few companies will be willing to perform research outside their classical markets, even if their activities are constrained in the ways mentioned above. Even fewer firms will be able to perform this type of research. Under the market conditions just mentioned, it is likely that price competition is fierce and profits are low. Then, shareholders and management are not likely to assign funds to research.

Third, the risks involved in performing research are usually considered to be higher than in other business ventures. This may or may not be true. What is important is the extent to which managers are willing to shoulder the risks involved. The less they are willing to do so, and the more riskless opportunities are offered and valued by the stakeholders of a company, the less we can expect them to support research.

Fourth, regardless of market conditions, a company may well leave it to others to engage in research, namely publicly sponsored institutions such as universities or research institutes. Results achieved in these institutions usually are made openly available, and it appears as if they could be picked up easily once they are considered valuable for the company. This seems to save money, and thus to improve the competitive position of the firm. Private resources originally devoted to research then could be saved or transferred, for instance, to development departments in business units, where the success of their employment is more readily visible.

Some companies prefer to take licenses from possible competitors or to imitate, and thus leave the research to others. Managers of such companies attribute weakening competitive positions of their research oriented rivals to their substantial basic research efforts. This is the argument of Mr. Schwarz-Schuette, CEO of Schwarz Pharma Corp., in an interview published in late 1996. He does not mention the institutional backing that producers of generics have enjoyed recently in many countries. This supports imitation strategies relatively more than innovation strategies. However, we will not study institutional or legal trends that could influence the relative strength of research-based innovation strategies.

Given these arguments, it is no great wonder that under the pressure of recent economic developments and the propagation of fashionable concepts like lean management, short development times, etc.

ionable concepts like lean management, short development times, etc. many firms tend to scale down their research activities. Interestingly, the recent trends of cutting back research expenditures seem only to be a reprise of earlier and similar developments*. Recalls the director of an aerospace laboratory: "For a time, decentralization proved to be an effective organizational solution to the problems of diversification and growth, but by the late 1950s there was a growing tendency within the divisions toward parochialism and interdivisional competition leading to redundancy and a concentration on short-run problems. Each division had its own research staff ..." (Ruedi, Lawrence, 1995, p. 508).

As business conditions change, the support for research seems to change as well. Political ideas and public opinion may be more or less favorable to industrial research. As both of these change over time, they leave impressions on managers and consequently on the level of support for research. This is nicely summarized in the following quotation: "In October 1929, President Hoover used the occasion of the Golden Jubilee for the incandescent lamp to broadcast a plea for more industrial laboratories and more industrial support for research. The days of the woodshed genius were past. It was the day of the well-equipped special research laboratory. Hoover spoke at what would turn out to be the pinnacle of fortune for industrial research before World War II. His terms of service ... helped to ensure that industrial research, particularly research done in cooperation with other research-performing institutions, was seen as public-spirited activity. The public mood would soon reject both that technocratic vision and its presidential advocate" (Graham, Pruitt, 1990, p. 183). F. D. Roosevelt won the presidential elections in 1932. During the 1996 presidential elections the Industrial Research Institute issued a message to the candidates asking them to "maintain the strong role of government in supporting undirected basic research" with further details regarding specific policy issues.

Similarly rapid policy changes can be cited for other countries as well. The fate of biotechnology and gene research in Germany during recent years serves as another example. It was first exiled to a large part by unfavorable public opinion and laws that put up heavy safety

---

* For instance, considerable interest in this question arose in the late 1950s. See for example: Furnas, 1958; Kinzel, 1961; Berthold, 1968.

constraints. More recently the attempts and the difficulties of trying to lure back research to the country from which it emigrated can be observed.

Public policy changes can be intensified by policy changes within firms. However, too many changes may be detrimental to the competitive position of firms. In a recent study of German companies, 75% of the economically weaker firms complain of the lack of steadiness in the objectives for research and development, while only 30% of the stronger firms voice the same complaint (Foos, 1995)*. In fact, changes may go back and forth, as companies lack good criteria and long enough planning horizons for evaluating and supporting their research activities. The history of DuPont demonstrates a pendulum swing from the relative strength in supporting research versus that of development, in part reflecting an impatience with the ability of the researchers to repeat past successes within a relatively short period of time (Hounshell, Smith, 1989). Such pendulum swings can also be the reflection of a search for an optimum organizational structure and optimum funding levels, where either 'hard' facts for its evaluation are lacking or evaluation is related to past observations that tend to fluctuate in response to changing market conditions. Another reason for missing the equipoise for research expenditures is that business planning uses past data to support its future directed investments. This is demonstrated in a case study concerning the early years of synthetic rubber research at Bayer, the German chemical company. The study indicates that price levels of natural rubber produced decisive signals to the management in their deliberations on whether to stop or to continue with this research (Brockhoff, 1996 b). In addition, recent trends favoring decentralization, at the same time disfavor research, as decentralized units tend to adopt more short term planning horizons than central units in firms.

Another interesting case for the demonstration of impatience with research and its consequences is reported by Peter Bridenbaugh, former Chief Technical Officer of Alcoa Corp. (Bridenbaugh, 1996, p. 155-163)[†]. It is worthwhile to quote his report at some length:

"The creation of the Aluminum Research Laboratories (ARL) in 1919 initiated formal R&D at Alcoa under the leadership of Dr.

---

* This refers to a study of 98 firms conducted by Arthur D. Little, Inc.
† The paper presents a personal, summarizing view of the study by M.B.W. Graham and B. H. Pruitt, 1990.

Francis Frary... By...1949, it was apparent that Dr. Frary and his colleagues had succeeded marvelously.

The next era began in 1950 and ended around 1965. It was sparked by domestic competition in the aluminum industry for the first time... In Alcoa's view the time had come to 'aluminize the world'. Application engineering became the major task for the R&D organization. The storehouse of knowledge built by Dr. Frary's organization became a resource to be guarded and exploited but not replenished. Engineers who had bachelor of science degrees were perfectly suited for the task at hand.

(In the 1970s) the company's clarion call became process improvement, and the major project of this era was a highly secretive, low-energy smelting process. Application engineering was dismantled as new product development was abandoned. If they thought about it at all, corporate leadership and isolated R&D management considered Frary's storehouse of scientific understanding to be sufficient to see Alcoa through. By the mid-1970s, it was apparent that Alcoa's R&D organization was failing to meet its objectives. The signals were vague at first but became clearer by the late 1970s, when a series of new corporate faces began showing up to 'fix' the labs. They succeeded in confirming the obvious but not in understanding the causes of the problem.

The next era began in the early 1980s and ... ended somewhere around 1988 or 1989. It was apparent that Alcoa had exhausted its foundation of scientific knowledge and had often engineered solutions to problems that were not fully understood. The highly touted Alcoa Smelting Process, for example, had failed. This failure was compounded by ... no meaningful new products. ... The leadership of the laboratory took it upon itself to define a new R&D charter that was not a great deal different from the charter given to Dr. Frary more than six decades before ... While corporate leaders never formally stated their agreement, they embraced these new directions not only with words but with dollars.

This brings us to the present era ... there is a significant portion of corporate management that thinks the initiatives of the 1980s were too ambitious and perhaps unnecessary. For these people, today is indeed a time of retrenchment and reduction in technical scope and effort. In the minds of others, however, it is a time to channel our knowledge and efforts into product and process development, just as the company did in the 1950s and 1960s. These people seek to harvest the knowledge that has developed in the last ten years."

We think that this rather bitter summary of technology policy in one corporation needs no further comment. In his summary of the developments the author unfortunately cannot reassure us that management has learned its lesson, but he only expresses hope that people "have learned that there is a price to be paid for abandoning the pursuit of fundamental understanding..." (Bridenbaugh, 1996, p. 159). In contrast, other technology officers pride themselves on a totally different idea: "Intel operates on the Noyes principle of minimum information: One guesses what the answer to a problem is and goes as far as one can in an heuristic way. If this does not solve the problem, one goes back and learns enough to try something else. Thus, rather than mount research efforts aimed at truly understanding problems and producing publishable technological solutions, Intel tries to get by with as little information as possible. To date, this approach has proved an effective means of moving technology along fairly rapidly" (Moore, 1996, p. 168)[*]. Certainly, there is no proof that a more systematic approach might have been even more successful, or to what degree the inflow of newly hired researchers from the best schools into the company may have secured the outcome that is described. The latter issue in particular may be critical. Alcoa stopped hiring research oriented personnel at some point in time, while Intel obviously did not. This could serve as an explanation for the fact that certain detrimental effects described for Alcoa have not arisen at Intel. Further explanations could be derived if differences in the characteristics of knowledge used in the new products or new processes of both firms were determined, particularly with respect to the degree to which one can learn from earlier results.

The Alcoa case also demonstrates that the weakening of competitiveness is a creeping process, the results of which are not immediately visible. This is characteristic of processes with long time lags between the outlay and the later generation of revenues, and with contributions to a stock or to potentials not used in the same period

---

[*] Interestingly, the procedure described here has some relationship with an integration model of the research process with other business functions that is called the 'chain-linked model'. It sees research to be immediately linked to practically every stage of the classical linear invention-innovation process. See: Kline, Rosenberg, 1986, here pp. 285 et seq.

in which they have been generated. Quite apart from the individual case presented here, similar conclusions can be drawn from simulation models of firm growth under competitive conditions. Outspending a competitor on research that promises to advance new generations of technology becomes clearly visible in higher sales and sales growth only after two generations of technology, and it is reflected in higher profits even later, due to the higher costs, but the effects are cumulative and tend to grow over time if the strategies at issue are retained (Weitzel, 1996, pp. 133 et seq.).

Fluctuating shares of research expenditures relative to total R&D expenditures within industrial firms of one country raise additional concerns. In the early 1980's, it was claimed that "basic science and intellectual capital seem certain to play a more fundamental role in relation to socio-economic development than hitherto. Recognizing this, high-technology companies have been channeling increased resources to basic research, especially in areas where fundamental scientific breakthroughs are needed if emerging generic technologies are to realize their full commercial potential" (Martin, Irvine, 1989, p. 1). It is ironic to observe that a book on research foresight could be misguided so dramatically, as quite different developments happened more recently in the U.S.: "As a result of the restructuring of many companies, the levels of their efforts in basic research have been attenuated. Their dependence on university research has increased. Industry has expanded its support of university research and entered into many hundreds of collaborative arrangements" (Abelson, 1995, p. 435). In the middle of the 1990's we notice that even highly reputed basic research laboratories in Europe and in the U.S. that can point at a phenomenal track record of research breakthroughs that have changed the world, have come under pressure to reorganize and to perform more applied research. One of the most prominent examples is the announcement of the reorganization of the Bell Laboratories of AT&T (Schmitt, 1995). The tendency reflected in such observations is of great concern among academic observers of technology management (Rosenbloom, Spencer, 1996).

Counter to these positions some firms argue - at least for the time being - that their own research is the major key to their success. Let us again look at a Japanese company as an example. Nichia Chemical Industries, a medium-sized company located in Anan, recently announced the production of a prototype semiconducting laser that emits blue light, which was not possible until then. This product was realized by 'niche research' based on experiences with indium-gallium-light diodes. The company expected that due to this research result it could expand sales tenfold from 1995 to 1996. It sees further potential markets if additional development works should prove to be successful (hra, 1996, p. 19). Of Reilly Industries Inc. it is reported that "it was the research laboratory that had the most influence on the company's evolution". One of the main reasons for its survival for one century is its "commitment to research ... that continued even through the company's lean years" as a consequence of the management's "unshakable optimism in the future of science" (Gwynne, 1996, pp. 39, 40). Numerous other examples support this experience.

Aside from proving the value of their research divisions, firms may also be forced to rely on their own research base if universities do not deliver the research results most desired and at the right point in time. This may have several reasons. At the strategic level, it may become dangerous in the long run if universities or publicly funded research institutions would follow influential advices to reduce their scientific orientation. This danger was suggested, for instance, with respect to engineering education in the U.S. (Dertouzos, Lester, Solow, 1989). This may reduce the production of research results needed by firms. At the tactical level, as well, it can be observed that university research does not completely fit industry needs. F. Betz of the National Science Foundation points to the long transfer times for university research results to industrial application, but also to a structural problem: "In universities, scientific research is integrated with graduate education, and projects are divided into 'thesis' size and duration. Much of the scientific research in universities is performed by doctoral candidates for their thesis requirements. Doctoral research is seldom at a scale, breadth, or timeliness suitable for industrial needs" (Betz, 1996, p. 1-8). While this view is a bit too one-sided, it has a true core. To overcome the structural problems, either other research institutions need to be established (which may then have their own problems with pressures to respond to industrial needs), or industry has to engage in its own research, if only to build a basis for joint industry-university research cooperations.

Thus, the crucial question put at the beginning of this chapter does not find an easy answer. Company management has little guidance in its decisions on whether to support research and at what level. In fact, finding a response has become even more difficult since the dramatic internationalization of competition. This leads to an even more complicated question: do companies behave differently in competing countries, and how should one react to the observed behavior?

This question is of particular interest with respect to the phenomenal growth of corporate research in Japan. It has been observed: "Unlike the United States, where basic research flourishes in academia with federal support, Japan lacks a strong public infrastructure for fundamental research. Stepping into the gap are Japan's major corporations, many of which have come to see building their own basic research as a necessary step in anticipating new technologies with commercial potential. Since 1985, just about every major electronics corporation in Japan has opened independent, and often somewhat freewheeling facilities in the suburbs of Tokyo that are devoted to fundamental studies in materials, computing, electronics, and, oddly enough, biology" (Hamilton, 1992, pp. 570, 571). In Table 1 we present some illustrative data on major private research laboratories in the Japanese electronics industry. It shows that this industry was comparatively late in engaging in research. Thus, the lack of a basic science tradition in public research institutions in Japan has been used to explain a growing share of private research expenditures in this country, which are to be used as a strategic weapon in competition (Martin, Irvine, 1989, p. 2). It was necessary to develop this strategic weapon when the imitation strategy suffered from its own success. But even in Japan some firms, such as Hitachi Ltd., have recently chosen to integrate parts of their research laboratories into business functions, in an attempt to tie them closer to business needs. As this move does not hit the whole research organization, it may also be interpreted as an indicator of the maturing of some of Hitachi's research into development. The organizational change could then be regarded as a demonstration of flexibility, particularly if new research units were to be set up.

This leads to yet other questions. Could the relevance and the accessibility of university research play a role in explaining differences in private research funding over time and between countries? Have companies in different countries developed particular strategies to access external research results?

It is difficult to draw conclusions from casual observations. The limited number of cases to which we have referred above make it clear that for each example one may also find a counter-example. Therefore, a more systematic and possibly statistically based discussion seems to be in order. We shall try to present this in the following chapters. In the second chapter we characterize research more systematically than was afforded by the remarks made above.

Table 1: Major basic research laboratories in the Japanese electronics industry

Laboratory	Date Opened	Number of Researchers	Annual Budget (mil. $)	Fields of Research
Canon Research Center	1985	250	n.a.	Optoelectronics, advanced materials, biotechnology
Hitachi Advanced Research Laboratory	1985	114	41	Electron beam physics, software, molecular biology
NTT Basic Research Laboratories	1985	200	25	Quantum optics, computer science, materials
NEC Fundamental Research Laboratories	1989	100	35	Advanced materials, atomic manipulation, neurobiology

Source: Science, 25 October 1992, p. 571.

Then, in the third chapter, we can derive reasons for performing research by private companies. We shall present evaluations of such reasons, such that the more or less important reasons become visible. From this discussion we can proceed to the fourth chapter, in which it is argued that research helps build certain potentials (or, as some management literature might prefer to say: it provides specific resource bases). Once these potentials have been identified, we can go on to use them for multiple purposes. They may provide a broader base for project funding, as shown in chapters 5 and 6, and they help us to identify the necessary conditions for research success. This is discussed in chapter 9. At this place we can also discuss some of the problems of integrating external research results into the firm. The necessary conditions need to be matched with sufficient conditions in order to achieve a basis for research success. We present some remarks regarding the sufficient conditions in chapter 8.

Research projects are aggregated to a research program. This bottom-up approach has to be balanced out by a top-down approach of budgeting that takes overall considerations of a strategic nature into account. There are at least two approaches to solving this task. Budgets are derived either from future business requirements or from present business possibilities. In both cases, one needs to have an idea of the potential contribution that research can make to generating future business. Here, the concept of elasticities comes into play. It is already widely used in areas such as marketing or manufacturing, but is not yet well understood in the research and development environment. In chapter 7 we demonstrate the strength of this concept, and we indicate approaches by which elasticities may be empirically determined.

Chapter 10 offers some remarks on the location decision for research departments within firms. We argue that the functions to be performed by a laboratory become influential in selecting its location.

In the concluding chapter we summarize some of the main observations and recommendations that can be drawn from this study for research management and for general management.

2. Characteristics of research

2.1 Problems of measurement and definition

An answer to the questions of whether and how much companies should invest to support research is much more difficult than is indicated by the few arguments mentioned above. In defense of in-house research, management may point to the relatively low cost of these activities. In fact, on average only a small share of the total research and development expenditure is devoted to research (see Figure 1). It is interesting to observe that this share has developed very differently in advanced industrial nations. It is stable and low in the U.S. until 1985, but after this year when data are collected differently it starts to reach higher levels and fluctuates considerably.

The share is higher and rising in Japan (at least as of the late 1970s; we suspect that the extremely high shares shown for earlier years are a result of unclear definitions or a biased application of the definitions in firms).

In Germany the share devoted to research shows substantial fluctuations over time. In fact, we observe a pro-cyclical relationship between the share of industrial research and the growth of the gross national product in the same year or two years ago. This could be a reflection of rather rigid, standardized, and past-oriented budgeting procedures. If so, this is not a healthy basis for the generation of contributions from research. The compound annual growth rate of real research expenditures in German industry between 1971 and 1991 was 3%, and it amounted to 7.4% in the U.S. In Japan this growth rate was 7.8% over the same period of time. The rate goes up to more than 13% if we consider only the period from 1975 to 1991, when more reliable data are available. In the three countries, real growth has declined and even became negative in recent years.

Fig. 1: Basic industrial research as a share of the total industrial R&D expenditure in three industrialized countries

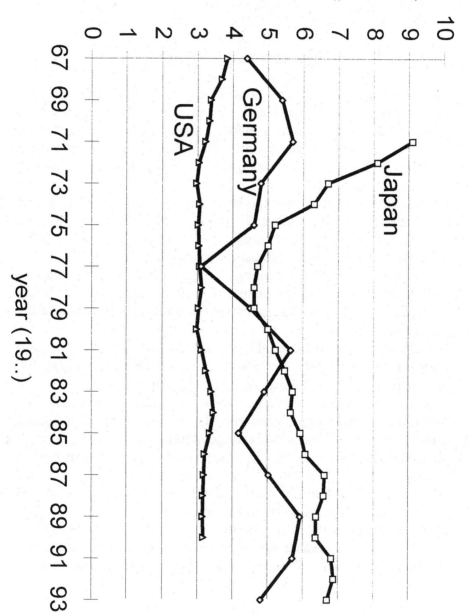

In Japan, increasing long term shares of basic research expenditures could reflect a strategic intent to conquer new market opportunities. This is a specific trait of Japanese management, as opposed to U.S. or European approaches that often aim at winning shares in established markets (Prahalad, 1993, pp. 41 et seq.). In this respect it is interesting to note that recent reductions in the share of industrial research expenditures coincide with a dramatic increase in the share of governmental basic research expenditures; a similar development was observed in the middle of the 1970s. This raises the question of whether the government acts deliberately to balance 'missing' industrial research in Japan (see the statistical data in the Appendix). A recently published report makes the government's position quite clear: "It is necessary for the government to actively implement R&D in fields the private sector does not undertake to study, such as basic and original research" (Government of Japan, 1996). A considerable number of activities have been initiated to this end.

Certainly, it is dangerous to argue with averages of research shares, as they are subject to more or less variance. Leading companies in the electrical and electronics industries spent about 10% to 12% of their sales on research and development. These median values have a variation from less than 5% to more than 20% during recent years. In 16 interviews with European firms in this industry, a median share of 9% for basic research of total research and development was reported; in 20 Japanese firms we have noticed a median share of 19% for the same relationship*. Differences in definitions of the term notwithstanding, the percentage shares represent large sums of money that - if saved - seem to raise the return on sales by one percentage point in the short term at least. Competitive pressure and impatient stockholders may exert pressure to reduce what appears to some of them to be an expenditure only, and to some others as being luxurious. Such short-term views are considered dangerous not only because they can undermine competitive positions in the long run but because they reduce potentials that could be used by further genera-

---

* Interviews by J. Bardenhewer for the project "Research in the Company of the Future" were conducted in 1995. On the basis of information taken from annual reports, H. Ernst shows an average share of 10.3% for 10 European firms in the same industry, and 18.5% for 15 Japanese firms. The differences are significant. The shares of total research and development, while being slightly higher in the European firms, are not significantly different. See: Ernst, H., 1996.

tions. Both aspects are convincingly demonstrated in the Alcoa case presented in the preceding chapter.

Research is a specifically chosen combination of resources aimed at generating new knowledge. In this respect, it may be indistinguishable from development. But while development is clearly aimed at immediately using the new knowledge in new or improved products or production processes, this is not expected of pure or basic research. Where the use of new knowledge for future new products or new processes is envisioned only rather broadly and not in very specific terms, the term of applied research often arises. In this respect, differences between research and development can indeed be recognized. The OECD (1992) builds its definitions of research and development on this observation. They are widely used in its member countries for preparing statistical information. Basic research is defined by the National Science Foundation (1959, p. 124) as "original investigation for the advancement of scientific knowledge ... which do[es] not have immediate commercial objectives". In this definition the word 'immediate' is of some interest. It implies that in a very long run basic research results may find an application. Thus, what appears to be 'basic' ex ante and in the short run, could turn out to be 'applied' ex post and in the long run (Agassi, 1966, p. 348). This involves an uneasy and only subjective reference to some kind of planning horizon. Similarly subjective intentions for research are addressed when it is asserted that basic research helps to satisfy a desire 'to know', while applied research applies its results to enable us 'to do' something (Feibleman, 1961, p. 305). Besides other problems this explanation builds implicitly on the nowadays discarded 'linear model' whereby basic research feeds applied research, and this in turn feeds technology.

In practice the definitions are not entirely clear and give rise to problems of demarcation. They also reflect dissimilar research or development traditions and cultures, manifest in the way that results from these activities are presented. Thus it can be shown that relatively more publications and patents result from research in Germany as compared with Japan (Foray, 1995, pp. 104 et seq.). It is argued that structural differences in property rights systems may have induced these differences. Such differences can have profound impact on the way that research and development activities are coordinated or integrated (Aoki, 1988, p. 247; Ordover, 1991, pp. 43-60).

The difficulties of definition have initiated a strong chorus of voices calling for more finely differentiated terms. However, this alone cannot entirely alleviate the problems. For instance, funding and guidance by top management, a long time-horizon, and the intention to support basic corporate purposes are considered to be additional characteristics of corporate research (Rosenbloom, Kantrow, 1982, p. 116). Unfortunately, as the same criteria can also become relevant objectives for development, they are not particularly distinctive.

From the perspective of university research that is no longer limited to pure basic research, it has been suggested that additional categories for a classification of research activities are needed, such as 'applied basic research' (Berthold, 1968, p. 146; Mittelstraß, 1995, pp. 18-24). If the apparent need for such changes in definitions signals changes in the character of research performed in universities, this would indeed tie in with the observation of industry's increased interest in collaborative research with universities, as mentioned above.

Unusually high shares of industrial research expenditures during the middle of the 1970s in Japan (see Figure 1) gave rise to the assumption that definitions were changed or interpreted very broadly. Research definitions and their interpretation can therefore have profound effects on the comparability of data that record research activities. This concerns time series data and cross sectional data as well.

The effect of a broader definition can be seen by comparing Figure 2 with Figure 1. The share of basic and applied research taken from total industrial R&D expenditures is about eight times higher in the U.S. and four to five times higher in Japan than those countries' shares of basic research alone. Similar observations can be made for individual companies. As research definitions have not been standardized, we have to expect substantial reliability and validity problems in measuring industrial research activities.

For the purpose of illustration of research activities at the company level (and as an additional indication of the growth of Japanese research activities), let us consider some organizational aspects of Hitachi's R&D. This company spends 10.1% of its sales on R&D (1993), which amounts to 3.7 billion $; 26% of this sum is spent in nine corporate research laboratories.

Fig. 2: Basic and applied industrial research as a share of industrial R&D expenditures in Japan and the U.S., 1981-1993

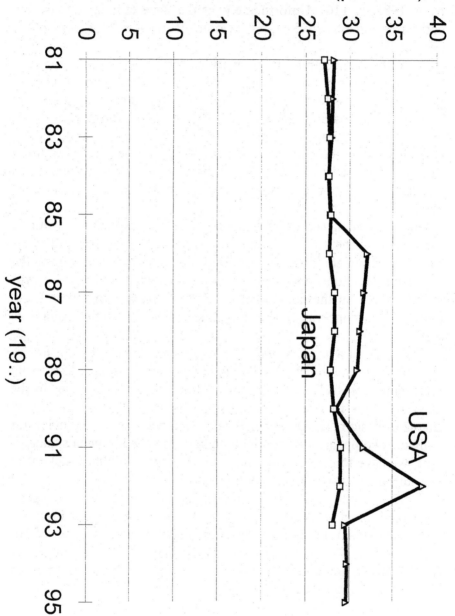

Since February 1995, two of these laboratories have been assigned to business units, such that the share of the corporate research laboratories was decreased to 17% of the total R&D expenditure.

Two of these laboratories (Central Research Laboratory, Advanced Research Laboratory) perform 'fundamental research', while the others carry out 'applied research'. In terms of number of employees, fundamental research claimed 25.8% (1993) of the staff of the corporate research laboratories. More recently, the budget share was reported to account for 8% of the total R&D expenditure. Again, this may indicate a drop in research efforts or a change of definitions or both.

The terms 'basic' and 'applied research' mentioned above could imply a hierarchy of information transfers. Pure research delivers results that are used by applied research, which in turn presents its findings to development. It has been made clear repeatedly that this 'linear model' of knowledge generation and transfer is not a completely correct representation of reality. Information could flow from any stage in such a model to any other stage, thus invalidating the linearity assumption. Still, the linear model could be used as a heuristic in examining the relationships between the different stages (Steinmueller, 1994, p. 55). It can be observed that at the same time 'basic' research may generate immediately applicable knowledge as well as profound insights that become relevant to applications only many years later. Thus, while there exists general agreement on some aspects of 'research' as separated from 'development', we cannot present a broadly accepted, operational definition.

2.2 Appropriability and transferability of research results

To defend privately supported research activities, a significant number of arguments must be considered, relating to
- the appropriability and the transferability of research results,
- the specific risk of research,
- the specific benefits that are expected from research.

We shall try to sketch these aspects very briefly.

New knowledge generates economic value to the individual firm only if the firm can effectively exclude other firms from using the same knowledge, for some time at least, and to some degree. In theory, research results can be used by an unlimited number of interested individuals or firms at no additional cost once they learn about these results. Therefore, firms tend to underinvest in research in a competitive situation. This argument was advanced by R. Nelson (1959). Rivals can be excluded from such free rider behavior only if the results can be protected legally, if they are kept secret or if the cost of transferring the results is high enough to exclude others from using them. However, patenting is only very rarely available for protecting research results, and keeping them secret constrains the company's own derivation of economic benefits from the newly gained knowledge[*]. Protection may not be perfect due to the cost associated with securing the protection, the cost of enforcement, of identifying infringements of one's own position and of stopping such infringements. Furthermore, a description of legally protected knowledge in a patent may provide clues for circumventing it.

The cost of transfer depends not only on one's own activities, but also on those of the potential users of new knowledge. Thus, exclusion of rivals may not be perfect.

Therefore, it is often and since long time thought that not firms but universities or other public research institutions "are the proper places for the pursuit of 'pure' science, and for the establishment of laboratories, etc., devoted to it" (Marshall, 1927, p. 100). If so, industry has little possibility of influencing the course of inquiry. In some countries, such as Germany, freedom of research is a constitutional right. Influence can be exerted only indirectly, through cooperation, funding or moral suasion.

Moral suasion can occur in general terms[†], or may take the specific form of suggestions for research topics. For example, the Association of the German Electrical Industry has produced a catalogue of such topics (ZVEI, 1994 a). Clearly, there are neither guarantees that moral suasion will have the desired effects nor that the fields of greatest interest to industry are supported by public money. A

---

[*] Some researchers observe new organizational developments in innovation processes that lead to more patent protection of research results. This would initiate a stronger need for close interaction between science and technology in individual firms. See: Foray, 1995, pp. 90 et seq.

[†] "...studies will lose nothing, and the world may gain much, from keeping in touch with some of the industries, whose methods might be improved by increased knowledge...": Marshall, 1927, p. 100. For a more recent example, see the reports on the publicly funded research centers in Germany: Weule, H., et al., 1994; ZVEI, 1994.

stimulation of certain types of university research could also be achieved by presenting fascinating first results of new phenomena. "It is clear that this sort of strategy cannot be pursued with second-rate staff", is the conclusion drawn from Bell Telephone Laboratories' behavior in the case of its successful stimulation of research into materials problems after the discovery of the transistor effect (Ruedi, Lawrence, 1995, p. 512).

Funding is a second-best solution from industry's point of view, given the above arguments on the appropriability of the benefits from the new knowledge. The level of different types of funding may well be related to the level of influence that industry may exert on the choice of research topics. Together with other alternatives this is shown schematically in Figure 3. Low funding shares and high influence are unlikely combinations, as are high funding shares and very little influence. However, a typical donors' behavior may come close to this situation. Sponsors tend to ask for a higher level of influence on the selection of research topics than donors. Donors and sponsors leave the actual research work to the people selected and employed by the recipient organization of their funding. This is not the case when companies enter research cooperations with industry. Here, personnel may be exchanged, visiting posts can be established, or industry representatives may serve on an advisory or a managing board of a separate research institution jointly operated by a university and one or more firms from industry. This offers many opportunities for influencing the choice of topics. In the case of one specific organizational arrangement, the so-called cooperative research centers, this fact was clearly spelled out: "Cooperative centers provide opportunities for university-industry interaction at each stage of the research process, i.e., in the planning of a research program, the design and implementation of research projects, and in the transfer of research findings to industry for development. The centers enable university and industrial researchers to coordinate their research agendas..." (U.S. General Accounting Office, 1983, p. 24). In the U.S., the number of such centers increased more than tenfold in only three years after 1989, and even the U.S. weapons laboratories have signed more than 300 cooperative research and development agreements with industry (National Science Board, 1996, pp. 4-28). This draws our attention at a further aspect. Introducing a shift of focus from defense related basic research to more commercially relevant basic research not only reveals that basic research can have a particular direction, but also shows that industry may need the government as a partner in initiating and forcing such changes (General Accounting Office, 1994).

An extreme case with respect to influencing the choice of topics is seen in contract basic research. Here, industry becomes a contractor if and only if it can come to an agreement with the university on the kind of work to be performed.

Fig. 3: Share of direct industrial research funding and level of industry's influence on the choice of topics in publicly funded research institutions, such as universities

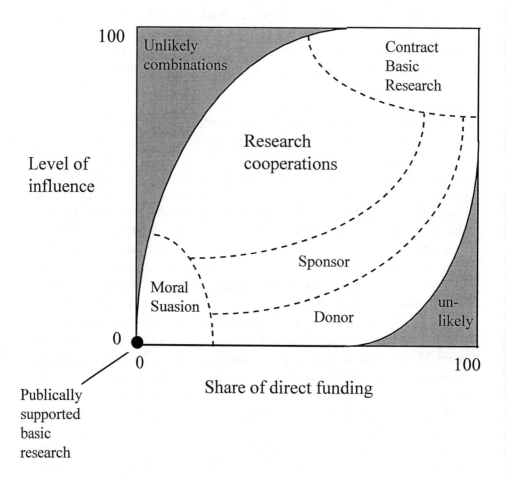

What may make things worse from the point of view of industrial technology managers, is that industry cannot always control the quality of publicly funded research. Establishing control may require contacts with the research establishment well beyond a particular project and, as far as international cooperations are concerned, also beyond national borders. It goes without saying that this is often quite difficult, particularly if foreign parties are granted only limited access to another nation's research institutions even beyond the hurdles erected by differences in language, culture, etc. Tapping foreign research resources may be desirable, if national research is not developed to a competitive level. To initiate this alternative may not be easy. Countries that feel deprived of their basic research results react by establishing institutional barriers to easy knowledge outflow (President's Commission on Industrial Competitiveness, 1985). Economically speaking, the cost of establishing access to the research results is higher for foreigners than for nationals. This cost needs to be added to the transfer cost that can be expected at the national level if industry decides to use results from research performed in other institutions.

In reality, even knowledge that is documented and available to others cannot be transferred without cost. This knowledge must first be identified; then, it needs to be absorbed by the internal organization, and it needs to be used. Identification and absorption require some research capacity in the company itself. The specific character of research that provides this capacity has been called 'absorptive capacity' (Cohen, Levinthal, 1990; Pisano, 1990). With particular reference to research cooperative centers, it has been said that a "firm's capacity to benefit from membership ... is heavily influenced by the sophistication of its R&D capabilities" (U.S. General Accounting Office, 1983, p. 30). This sophistication is a necessary condition for the development of a two-way information exchange between the industrial laboratory and its research environment as well as the fostering of high-attractiveness for creative researchers. These developments can be a source of conflict between researchers and managers in an industrial environment (Graham, Pruitt, 1990, pp. 153, 204).

It is apparent that in many companies research has to win or to re-win the support of higher management levels for itself and its interaction with the scientific community. Here we may quote an experienced chief technology officer: "We need to educate senior management, even CEOs, about the importance of a strong technology base and how to leverage it (and) about the symbiotic relationship between basic research and industry-directed fundamental work" (Bridenbaugh, 1996, p. 162). Unfortunately, this symbiotic relationship can survive only if both types of institutions adhere to implicit rules. Industry needs to perform some research, and universities or similar institutions, too, continue to provide a major share of basic research. If the latter institutions - either for ideological reasons or due to budget constraints - would change their mission, such as to perform exclusively development work or application engineering on a contract basis for industry, there can be no counterpart for industry to nurture and to enjoy the benefits of a symbiosis. To secure the interplay between institutions in the manner put forth by Bridenbaugh in the above quotation, it appears that a systems' view of the knowledge generating institutions in a nation is necessary if a terrible trap is to be avoided: the trap of the common lowest level of R&D. As early as 1945, the United States Office of Research and Development stated in its "Report of the Committee on Science and Public Welfare" that it "is important to emphasize that there is a perverse law governing research: Under the pressure for immediate results, and unless deliberate policies are set up to guard against this, applied research invariably drives out pure. The moral is clear: It is pure research which deserves and requires special protection and specially assured support" (Steinmueller, 1996, p. 56). This was addressed to government, but a similar message may be sent to the general management of many companies.

While the ideological reasons for the development of such a trap have widely disappeared with the weakening of socialist ideas that would have favored them, the freezing or cutting of budgets for governmentally supported institutions of learning and research in many countries has become a severe danger. In some instances it has come to the point that the demands of industry as partners in cooperations with universities have erected the major dams against an ever eroding basic research orientation in universities*.

The industry-university division of labor in the production of research and development results is not only a matter of quantitative

---

* In Germany, it is notable that even a daily newspaper runs a report mentioning the differences between fund-stripped universities and Max Planck Institutes that, for the time being, can afford to concentrate on pure research. Rubner, 1996, p. 10.

effects, as it may appear from the above arguments. The more universities lean towards development jobs for industry, the more relevant their results are for immediate use and very much targeted to influence competitive positions. This can counteract the ethical standards of university research that include unconstrained publishing of results. Today, universities already have agreed to publication restrictions in their contract research contracts, and at least one company has exercised its right to deny publication of unfavorable research results (King, 1996, p. 1). With an increasing share of industry-sponsored research, universities could become more susceptible to agreements to such standards and to the resulting behavior. Certainly, the same danger arises if individual government agencies tie their financing of university research to comparable contracts and monopolize the resulting knowledge in a non-benevolent way. However, we will not advance this group of problems any further here.

Use of externally generated knowledge can be achieved only if the 'not-invented-here'-syndrome does not stand in its way. The syndrome can be modeled as a transfer cost in economic terms. It can be shown that increasing cost of transferring new knowledge into a company reduces the use of external knowledge, but increases the use of internal knowledge up to a certain level only, as this broadens the basis for the internal integration of the external knowledge. Increasing strength of the 'not-invented-here'-syndrome reduces the use of both internal _and_ external new knowledge (Brockhoff, 1995, pp. 27-42).

To summarize and to make things crystal clear, Pavitt argues that even though "results of science are a 'public good' (i.e. codified, published, easily reproduced and therefore deserving of public subsidy)" it would be unreasonable to assume "that they are a 'free good' (i.e. costless to apply as a technology, once read)" (Pavitt, 1991, p. 112). This holds an important message not only for the potential use of external knowledge but also for the support of industrial research itself. Either by nature of the new knowledge or by choice of arrangements at least some degree of appropriability of benefits from new knowledge can be achieved. This is an important impulse for industrial research either for knowledge generation or for knowledge acquisition.

2.3 Specific risks of research

It is generally asserted that research is more risky than development (Brockhoff, 1973, pp. 28 et seq.). This does not mean that the probability of completing a research project successfully is much different from that of completing a development project. However, given the very loose relationship between the knowledge derived from research and its use in products or processes as compared with the much tighter coupling in the case of the development knowledge, the probability of achieving a return on the investment for a particular research project may be lower than for an equal investment in a particular development project.

Returning to company in-house research, it should be kept in mind that planning procedures require to suggest in advance (ex ante) whether the support of general corporate purposes can be achieved by such activities. This requires the demonstration of a link between present inputs and future results. To prove this point, a reconstruction of the contributions of new knowledge to product or process innovations is often attempted in retrospect. This is indeed very fascinating. A particularly interesting example of such studies is a report prepared for the National Science Foundation on the antecedents in science of revolutionary new products, such as magnetic ferrites, the video tape recorder, the electron microscope, the oral contraceptive, and matrix isolation (IIT Research Institute, 1968, 1969). Similarly, the Naval Research Advisory Committee initiated a report outlining the beneficial activities carried out by the physical scientist I.I. Rabi and his group of basic researchers to the U.S. Navy (Arthur D. Little, Inc., 1959). In more general terms, Alfred Marshall observed some 70 years ago: "History shows that almost every scientific discovery, which has ultimately revolutionized methods of industry, has been made in the pursuit of knowledge for its own sake, without direct aim of any particular practical advantage" (Marshall, 1927, p. 100). The General Electric central research laboratory estimates what the company would have missed out on, had the laboratory not existed, and uses this result as one of many yardsticks for research evaluation (Robb, 1991).

It is a different question whether such advances and contributions can be forecasted and planned. High uncertainty and long lead times "make for a poor fit with most planning systems", and managers lack capabilities to "target research clearly on a specific product or ... on a specific field of business" (Rosenbloom, Kantrow, 1982, p. 118).

Even favorable experiences with research within one company do not indicate future successes and thus cannot be projected into the future. This is very impressively described in the case of Du Pont. After Carother's success of discovering Nylon in his research laboratory, the company supported research and hoped for "new nylons" that did not appear upon command (Hounshell, Smith, 1989, p. 383). When impatience won the upper hand, development again received relatively more support than research. The company's pattern of support for research versus development follows an almost pendulum-like movement, seemingly accented by the most recent experiences of where new knowledge has originated. A similar problem was faced by Howard Schneiderman, Monsanto's senior vice president for research and development. His belief that "out of great science will come unique insights that will lead to product opportunity" was countered by strong arguments that "all research should be parceled out to operating divisions: this would have created market focus at the expense of central research capability. Corporate Research had as yet produced no biotechnology products and the ones under development ... were several years away from commercialization" (Leonard-Barton, Pisano, 1993).

Relatively higher risks of research would not be a problem for research support and indeed favor it if the majority of managers were risk-seeking. Experimental research as well as observations of managerial behavior indicate that most managers make risk-averse decisions, particularly if a company is profitable (MacCrimmon, Wehrung, 1986). It should be noted that risk-aversion does not mean that managers would never invest in risky prospects. It only means that if managers can choose between a certain return of 100 and a lottery that offers nothing or 200 with equal probabilities they are most likely to prefer the certain return. The risk attitude is reflected in the utility function that is explicitly or implicitly used to evaluate the outcomes of risky investments. The more risk-averse a manager is, the less he accepts investment opportunities that have large variances of returns. Even if the expected net present value of a research project could be determined at, say, 100 and that of a development project at 80, the risk-reverse manager may reject the research project if the variance attached to it is, say, 40 while the variance of the development project is only 20. For risk-averse people, increasing variance must be more than compensated by increasing expected returns.

Obviously, risk-averse behavior can eliminate high-risk research, even if its expected returns may let it look promising. According to a report in Business Week, the contact-lens manufacturer Bausch & Lomb appears to have lost its technological and economic lead in the beginning of the 1980's because it placed too much emphasis on less risky product improvements during times of affluence (Business Week, 1987).

2.4 Specific benefits from research

Observing strong environmental turbulence could trigger a reaction by which every activity is eliminated that does not immediately strengthen the short term competitive position of a firm. We have already indicated that this is short-sighted. Three more general thoughts are offered to strengthen our view: (1) New technologies reduce the dependency on random or chance events that spoil one's efforts to improve the quality of life. If the impact of such events grows, the potential contribution of research should be valued higher. (2) If turbulence in established markets increases, it attracts more attention by managers. Thus, attention to develop new opportunities is reduced if managerial capacity is limited. This may help new entrants to fight the incumbents. If the incumbents want to survive, they should deliberately allocate an appropriate amount of resources to strengthen future competitiveness. Otherwise, corrective action may come too late. (3) If market revolutions would truly become a regularly recurring event, then they could be anticipated, their consequences could be calculated, and research to bolster the consequences could be initiated. Thus, increased turbulence should in fact be an invitation to harvest the specific benefits that research can offer.

Generating income from new knowledge that could be used in new products and new processes is but one of the specific benefits that can be expected from research. Mansfield (1991, Tab. 1) found that 11% of the new products introduced by large U.S. firms during the decade from 1975 to 1985 could not have been developed (without substantial delay) in the absence of recent academic research, and another 8% relied on substantial support from this type of research (see also: Robb, 1991). Two percentage points less are reported for processes. Recency is defined here as a period of 15 years preceding the respective innovation. If these figures are considered reliable and representative for U.S. industry, corporate basic research should then be expected to have provided at least the same level of input for innovations. Unfortunately, it takes an average of seven years to transfer academic research findings to their first commercial use (Mansfield, 1991, Tab. 3). Again, this can be expected to occur faster within a firm if it has developed the necessary capabilities that support such a transfer of knowledge. Eggers (1997, Tab. 6, C2) has found the time span to be 6.4 years for German firms and 5 years for U.S. firms. The average 'research program time' of U.S. central laboratories is 3.75 years, including one year spent in technology transfer (Bosomworth, Sage, 1995, p. 36). At the end of this time the technology is not yet marketable, and the authors suggest that between 1.5 and 3 years may need to be added for a product to arrive at the market. Such time lags reduce the value of otherwise profitable research. To compensate for this effect, firms become interested in other potential benefits from research.

Given the specific character of research and its high risk, supporting the development of new products or new processes may not be the only benefit companies expect to derive from this activity. This argument should not come as a complete surprise, since it has been used by research directors for quite a number of years. As far back as December 18, 1926, director Stine of DuPont sent a memorandum to the executive committee of his company spelling out more reasons why research needed to be supported: "First was the scientific prestige or advertising value to be gained through the presentation and publishing of papers. Second, interesting scientific research would improve morale and make the recruiting of Ph.D. chemists easier. Third, the results of DuPont's pure science work could be used to barter for information about research in other institutions. Fourth, pure science work might give rise to practical applications. Although Stine personally believed that these would inevitably result, he felt that this proposal was totally justified by the first three reasons" (Hounshell, Smith, 1989, p. 223).

The spectrum of reasons may explain why, on aggregate, companies that perform research appear to enjoy higher growth rates and stronger productivity increases than other firms in the same industries. While the statistical analyses supporting this view (Mansfield, 1980; Link, 1981; Griliches, 1986) may be criticized in many respects (Brockhoff, 1994, p. 77), the underlying result does disappear. In only one study is it asserted that the R&D and capital expenditures of some large U.S. high-technology firms exceed their value to the shareholders during the 1980s (Jensen, 1993).

Two important issues arise from this view. First, it would be interesting to learn whether the enumeration of reasons for research as given by Stine is exhaustive or not. Second, it would be interesting to discover the ways in which corporate planning responds to such a diversity of reasons particularly as most of these point to benefits that are difficult to evaluate, to say the least. We shall deal with these questions subsequently.

3. Reasons for research

It is clear from the memorandum by Stine that research serves more than one function. In another early study on basic research in industry, Berthold (1968, pp. 175 et seq.) collects quotations from U.S. and German research managers who stress the importance of keeping pace with their scientific environment and to make good use of its results, on top of developing innovative ideas. Nokia AB describes ten functions for Nokia Central Research:
"Explore and develop
1. new technologies and their innovative applications and solutions for products,
2. new system and product concepts based on new or emerging technologies, including those falling between/outside the scope of current business units,
3. international patent rights and inputs to key standardization activities,
4. methods, tools and process know-how for enhancing the speed, productivity and manageability of the business units' products / processes,
5. key 'next best' alternatives and 'second opinions'.
6. Offer business units' product development subcontracting and consultancy by providing means for technology transfer, leverage competencies in critical product development tasks, being a vehicle for interaction that builds mutual understanding and trust.
7. Provide an environment for the exploration of new business opportunities (including those falling outside the scope of current business units).
8. Present insight and learning on new technologies for the business units.
9. Provide skilled personnel to business units' R&D-units.

10. Manage Nokia's interface to international R&D cooperations"*.

This is a very demanding list, and it raises the question of the degree to which it might be considered generic. Therefore, attempts were made to identify these functions more systematically. Members of a working group of the European Industrial Research Management Association pointed to six reasons for performing basic research in industry:

1. It leads to new developments,
2. it helps in understanding of processes and products,
3. it is necessary to remain informed,
4. it maintains scientific and technological standards,
5. it motivates researchers,
6. it attracts researchers (EIRMA, 1982, p. 12).

This list reflects the ranking order of importance of the items. Item 4 may have two different meanings: here, it is interpreted as firms being enabled by their basic research to conform better to standards that may be set by "society" and that are communicated and perhaps even multiplied by the media. This is the only item in the list that was not addressed in the empirical studies that follow. The long lead times of basic research will most likely reduce its applicability in presenting quick solutions to problems that might be voiced by society and media. At Alcoa, the rationale for establishing the first research laboratory was "the need for an independent technical authority to set corporate product and process standards and to mediate in the increasingly acrimonious disputes between the different Alcoa works over the quality of the materials that passed between them" (Graham, Pruitt, 1995, pp. 73, 101).

The ten functions presented in Figure 4 were identified on the basis of case research (Rosenbloom, Kantrow, 1982).

On the one hand, it is interesting to note that some functions, particularly those providing corporate services, do not follow from the specific characteristics of research as identified above. The question remains, whether further functions might be identified. On the other hand, some other functions following from the characteristics outlined above remain unmentioned. Furthermore, no information is available as to the importance of these functions, except scattered evidence collected in a limited number of interviews.

---

* From interviews, June 6, 1995.

Interviews with German R&D managers have elicited a list of 12 functions that have substantial overlap with the functions mentioned in Figure 4, and they point at some new functions as well. Presenting these to a sample of 26 research managers from large German firms, it was possible to rate the relevance of these functions. Results are presented in Table 2.

In Table 3 we compare interview results of German and U.S. managers. It is shown that - in spite of some noted differences - there is significant positive correlation between the responses.

High levels of consent for the first three functions mentioned in Table 2 with low standard deviations indicate that research managers consider the innovation-supporting functions of research to be more relevant than the service-related functions. Within this group, functions that support technology transfer are given higher relevance scores than those facilitating human resource management, image improvements or attracting public funding.

Fig. 4: Functions of corporate research

		New strategic directions	Supporting existing businesses
Innovation by	Improving and strengthening understanding of technologies in use	Corporate diversification to new applications and markets	Product and process improvements
	Discovering and developing new technologies	Corporate diversification to entirely new businesses	New processes for established products
Corporate service by	Intelligence	Windows on new science and technology	Assessing threats and opportunities
	Human resources	Recruiting new kinds of skills	Recruiting talented people with high potential
	Technology transfer	Identifying acquisition candidates with needed technological expertise	From corporate research to operations

Source: Rosenbloom, R.S., Kantrow, 1982., p. 120.

Table 2: Functions of research in 26 German companies
(scale values from 0 = does not apply at all, to 6 = completely correct)

Our own research ...	Mean	Std.dev.
... is a source of innovations (1)	4.62	1.60
... helps us understand the technology of our existing products or processes better (2)	4.27	1.40
... helps us improve existing products and processes (3)	4.15	1.46
... improves relationships to universities and other research institutions (4)	3.85	1.67
... increases our alertness vis-à-vis new developments in science (5)	3.85	1.71
... simplifies the application of research results from universities or other research institutions (6)	3.73	1.61
... simplifies the evaluation of research results from universities or other research institutions (7)	3.62	1.98
... simplifies the acquisition of scientific know how and methods (8)	3.23	1.70
... improves our image (9)	2.96	1.87
... supports the hiring of new research personnel (10)	2.38	1.81
... is a by-product of our applied research and development (11)	1.96	1.97
... is pursued because it gets public funding (12)	1.00	1.33

Source: Eggers, 1997.

Table 3: Comparison of German and U.S. perceptions of research functions

Our own research ...	Mean		Ranks	
	German	U.S.	German	U.S.
...is a source of innovations	4.62	3.75	1	2
...helps us understand the technology of our existing products or processes better	4.27	3.88	2	1
...helps us improve existing products and processes	4.15	3.63	3	3
...improves relationships to universities and other research institutions	3.85	2.50	4	10
...increases our alertness vis-à-vis new developments in science	3.85	3.00	5	5.5
...simplifies the application of research results from universities or other research institutions	3.73	3.00	6	5.5
...simplifies the evaluation of research results from universities or other research institutions	3.62	2.88	7	7
...simplifies the acquisition of scientific know how and methods	3.23	3.12	8	4
...improves our image	2.96	2.13	9	11.5
...supports the hiring of new research personnel	2.38	2.15	10	8
...is a by-product of our applied research and development	1.96	2.63	11	9
...is pursued because it gets public funding	1.00	2.13	12	11.5
Rank correlation: 0.74 ; $p < 0.05$				

Source: Eggers, 1997.

The impression given in Table 2 might be biased by high correlations among some functions. Therefore, it was tried to identify groups of functions that are largely independent of each other*. Three factors were identified:

(A) Research as a source of innovations (related to the functions (1) and - negatively - to (11)); this is meant to assure long-term competitive advantages;

(B) Research as a source of improvements (related to the functions (2) and (3)); this is meant to maintain short-term competitiveness;

(C) Research as a service to the company (related to all other functions) (Eggers, 1997).

Firms from the electrical and electronics industries attribute higher relevance to factor A than to factor B, which is then followed by factor C. Still, even the service-related factor is not irrelevant. However, it is very rarely used as an argument to support research activities, although some of its important ingredients were already mentioned in Stine's memorandum to the DuPont directors.

Evaluations of the present relevance of research functions may be different from their perceived ideal levels. In yet another interview study, we asked 17 research directors of major companies in the electrical and electronics industries in Europe (the majority being from Germany) and 21 in Japan to evaluate the present relevance of research functions along with ideal levels of the same functions. European and Japanese responses differ significantly only with respect to one item: The ability of research to provide a knowledge base for future development efforts, which would then lead to new products or processes. As far as the ideal case is considered, European managers assign higher ratings to this variable than Japanese research managers. With respect to the interpretation given for factor (B), above, one may say that European managers are relatively more interested in maintaining short-term competitiveness. Because of the high degree of agreement, we have decided not to differentiate the responses by the continent of origin. The results of these overall evaluations are given in Table 4.

---

* This was achieved by factor analyses, using the Eigenvalue-criterion for the selection of factors and Cronbach's Alpha as an indicator of their coherence.

Table 4: Actual and ideal performance of research functions as seen by 17 research managers in Europe and 20 in Japan (scale values 0 to 6 as in Table 2)

Functions	Actual		Ideal		Sig.
	Mean	Std.d.	Mean	Std.d.	
Provide a knowledge base for future development efforts - prepare for future products or processes	4.4	1.1	5.3	0.8	**
Provide information on emerging technologies to development	4.4	1.0	5.1	1.1	**
Provide information on emerging technologies to top decision makers	3.9	1.4	5.1	1.3	**
Observe competitive actions	3.5	1.2	4.7	1.1	**
Provide a knowledge base for easier implementation of external research results	3.5	1.2	4.2	1.1	**
Provide attractiveness to external research personnel	3.3	1.5	3.9	1.6	n.s.
Support the image of a company that contributes to technological progress	3.3	1.4	3.8	1.6	n.s.
Ease training of development staff in new skills and techniques	3.2	1.4	3.8	1.4	*
Socialize new staff members with the company culture for later transfer to other departments	2.7	1.3	3.5	1.5	**

Source: Interviews by J. Bardenhewer, Company of the Future Project, Kiel 1995.
Std.d.= Standard deviation; Sig.= Level of significance, with n.s. = not significant, ** = significant at a 99%-level, * = significant at a 95%-level.

All ideal states are rated higher than the actual states. Neglecting tied values, the rank orders of the responses are identical. Managers want to give each function more weight, but without changing their present rank orders. Interviewees were not forced to make trade-offs in their responses. The smallest differences in weights are observed for those research functions that support image-building and that increase attractiveness to external researchers. These differences are not significant. All other differences are significant. The largest difference is observed with respect to information flow on emerging technologies to top decision makers.

This could reflect the observation that the top management of many companies has developed a critical view of research in the company during the recent periods when extreme competitive pressure was put on their firms. Research managers seem to think that delivering better services to the top management might improve its evaluation of internal research. This phenomenon may also reflect growing recognition of the fact that one technology might affect many of the traditional industries. This becomes apparent from the statistics reproduced in Table 5.

The differences between the observations on actual and ideal performance of research functions can be summarized by factor analysis. This leads to three mutually independent directions for improvement:

(D) Supporting development efforts. This is achieved by providing information on emerging technologies and a knowledge base. It involves the two most important variables in Table 4.

(E) Supporting external visibility of the company. Supporting the company image, creating an attractiveness to external researchers, and providing a knowledge base for the easy implementation of external research results contribute to this factor.

(F) Servicing. The third factor draws on two different components. On the one hand it is made up of the socializing of new staff members and the training of development staff. On the other hand, it is related to technological competitor intelligence analysis and the provision of an information base on emerging technologies to top management.

Table 5: The relevance of scientific fields to technology in the U.S.

Field of science	Number of industries (out of 130) ranking scientific field as highly relevant (5 to 7 on a 7-point scale) with respect to their	
	Skills	Knowledge
Material Science	99	29
Computer Science	79	34
Chemistry	74	19
Metallurgy	60	21
Physics	44	4
Applied Math & Oper.Res.	32	16
Mathematics	30	5
Agricultural Science	16	17
Biology	14	12
Medical Science	8	7
Geology	4	0

Source: Nelson, 1987.

Factor (D) indicates a support of innovations similar to the factors (A) and (B) that were identified earlier. Factors (E) and (F) point to different aspects of services that may be provided by research and that are similar to the factor (C) that was found above. These factors indicate quite clearly that management is well aware that downstream-coupling of academic research, corporate research, and development is a necessary condition to achieving returns on their own research. This factor was widely overlooked in the economics literature until very recently (Pavitt, 1991).

The functions evaluated in Table 4 are not fully identical with those in Table 2. This is demonstrated in Figure 5. The figure might show more overlap between the studies than is actually there, because we choose expressions for the items that are thought to reflect their meaning but not necessarily their precise operationalization. Still, we can see some common underlying features. This is further supported by the similarity of the factors that were identified. In the following we shall try to integrate such factors into a concept of 'potentials' that research could provide.

Fig. 5: A comparative overview of studies on functions of industrial research

Function	Stine, 1926	Berthold, 1968	EIRMA, 1982	Rosenbloom, Kantrow, 1982	Nokia, 1995	Eggers, 1997	Bardenhewer, 1996
Better understanding presently used techniques			*	*		*	
Informing top management on new technologies				*	*	*	*
Informing development on new technologies				*	*	*	*
Competitor technology intelligence		*					*
Improving contacts with universities (or other public research organizations)		*				*	
Transferring technologies from external sources into the firm		*		*		*	*
Improving existing technologies						*	
Finding new technologies				*			
Being a source for innovations	*	*	*	*	*	*	*
Finding new systems and product concepts					*		
Facilitating personnel acquisition	*		*	*		*	*
Human resource development and motivation	*		*		*	*	*
International patenting and standardization					*		
Establishing (international) research cooperations					*		
Image improvement	*						*
Attracting public research funding						*	
Side-effect of development						*	
Setting and observing internal quality standards			*			*	
Strengthening information exchange	*					*	
Transfer from research to development				*	*		*

Sources: Stine, 1926: from Hounshell, Smith (1988); Berthold (1968); EIRMA (1982); Rosenbloom, Kantrow (1982); Nokia from interviews conducted in 1995; Eggers (1997); Bardenhewer (1996).

4. Research as a source of potentials

It is clear from the foregoing data that research managers agree that research performs not just one, but many functions within the company, and that most of these functions need to be strengthened. This is particularly true for the so-called service-related functions.

The functions are not performed for their own sake. Rather, they help to develop potentials that could be used to strengthen competitiveness*. Drawing on the characteristics of research as well as on the empirical studies in the preceding chapter, we can partition these potentialities into primary and secondary potentials. Later we argue that particular combinations represent necessary conditions for research success. Secondary potentials may contribute to success in a less stringent form. If the necessary conditions for research success are met by sufficient conditions this determines success, and it influences resource allocation among projects which if successfully completed support research functions. A sketch of this concept is presented in Figure 6.

The following primary potentialities are presented in the order of an idealized research process:

(a) *Identification potential*. This enables the company to identify technological knowledge that may become relevant for its present or future products and processes. In short, this may be called relevance for entrepreneurial use. The importance of this potential has already been exemplified by the data presented in Table 5. Recognizing the identification potential has a long tradition in economics. It was Adam Smith who observed in 1776: "Many improvements have been made by the makers of machines ... and some by ... those who are called philosophers or men of speculation, whose trade is not to do anything, but to observe everything; and who ... are often capable of

---

* Here, we will not discuss the various concepts of potentials. These are seen in relation to competitive strategies as well as to their resource base, which is closest to the view adopted here, or the development of core competencies. For a broadly based presentation of the different concepts see: Binder, Kantowsky, 1996 pp. 19 et seq., 43 et seq.

combining together the powers of the most distant and dissimilar objects. In the progress of society, philosophy and speculation becomes, like every other employment, the principal or sole trade of a particular class of citizens" (Smith, 1910, pp. 9 et seq.).

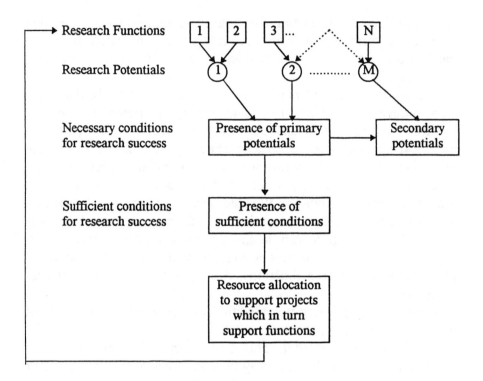

Fig. 6: A sketch of our concept

Requesting the employment of researchers who specialize in identifying external science and technology by referring to a more than 200 year old text may seem far-fetched to some; therefore, we supplement this request by some more recent testimonials. In one case study, a research director has remarked: "Many modern developments are so specialized that they can only be interpreted by someone who has a working familiarity in a new field. Thus, it may sometimes be of value to support a small research effort in order to keep an eye on the new field" (Ruedi, Lawrence, 1995, p. 512). The head of Hitachi's R&D, Yasutsugu Takeda, says in one of his contributions: "In order to discover the seeds and ideas for industrial innovation outside the company, especially in academic societies, the company should retain selected top-level researchers, whose activities are highly regarded in the world" (Takeda, 1996). Augustus Kinzel, a former Vice President of Research at Union Carbide Corp., points to the fact that in most cases much more research is performed outside the company than within; companies would have to engage in their own research simply to understand what is going on outside (Kinzel, 1961). Many more authors support the views expressed here (Berthold, 1968, p. 177). One of them chose to formulate it as follows: " ... the performance of basic research may be thought of as a ticket of admission to an information network ... The most effective way to remain effectively plugged in to the scientific network is to be a participant in the research process" (Rosenberg, 1990, pp. 170 et seq.).

(b) *Absorptive potential*. If interesting technological or scientific knowledge has been identified outside a company, it then needs to be integrated with internal research efforts: "If we will just allow ourselves the luxury of dispassionate observation, we will see the all-too-imminent chasm facing any corporation that is technically isolated and incapable of harnessing scientific progress because it cannot recognize and apply it" (Bridenbaugh, 1996, p. 159).

The potential considered here, thus describes the ability to absorb external technological knowledge for possible entrepreneurial use. In other words, the potential enables or facilitates the transfer of new knowledge from external sources into the company. A strong 'not-invented-here'-syndrome reduces this potential. As was observed by one of our interviewees: "My experience is that to exploit external research more effectively requires some internal research in the relevant field. Keeping up with science in the library, as a secondary activity for a researcher, is ineffective. I expect these trends to intensify." As argued above, this is a necessary condition for information exchange with other scientists. While this is common understanding, it has also been established empirically for the field of development (Schrader, Sattler, 1993). There is no compelling reason to assume that the same behavioral conditions could not be transferred to the exchange of information on research. In earlier years, this was described as active participation in an 'invisible college' (Price, 1963, pp. 62-91; Berthold, 1968, pp. 129, 183 et seq.).

(c) *Creative potential*. If new external technological knowledge has been identified and transferred into the company, researchers need to make good use of it. Also, creative ideas result from their own efforts. The ability to develop new knowledge is described by the creative potential. Again, a strong 'not-invented-here'-syndrome is a barrier to the creative potential. The creative potential may be used to support new business or existing business: "Such changes of emphasis are cyclic and are a response to prevailing business climate. For example, in the late seventies and early eighties, the emphasis ... was on generating ideas for new business. Now it is almost entirely on bringing in new science and technology in support of existing business", reports an interviewee. This is not the place to present and discuss institutional, group related, and individual conditions for increased creativity and for a goal-related controlling of creativity (Hauschildt, 1993, Chap. 9, with more literature).

(d) *Interpretive potential.* This potential describes the ability to evaluate existing technologies and to understand them better. It is not uncommon that some techniques are known but the principles behind them are not. Were these principles known, the techniques could then be employed more reliably and perhaps on a much broader basis. This would support competitiveness. Take ceramics for example. The art of pottery has been practiced for thousands of years and continuously been improved through learning-by-doing. It is only recently that the relevant physics and chemistry could be explained satisfactorily. Steam engines were constructed long before basic principles of thermodynamics had been established, such that improvements had to be made on the basis of learning-by-doing and observation of experience. A much better understanding and a higher learning rate could have been achieved by research, had this been initiated earlier.

(e) *Internal transfer potential.* Few research results can be applied immediately in new products or new processes. Very often they must be adopted by means of more or less elaborate development processes. The transfer potential describes the ability of research to create favorable, necessary conditions for a knowledge transfer into development groups (or other 'downstream' departments). An interesting case that illustrates the consequences of a lack of the internal transfer potential is that of Fairchild Semiconductor: "As it was in the right technology and functioned quite effectively, the laboratory was highly productive for a time. Then, in the late 1960s, it began to have difficulty transferring new products and technology to the product and production divisions. In 1968, for example, Fairchild still had not transferred to production MOS transistor devices that it had had stable in the laboratory since 1961 even though the technology was being exploited successfully by companies Fairchild had spun off and by spin-offs of those spin-offs" (Moore, 1996, p. 167).

Statistics indicate considerable internal transfer problems that may be attributed at least in part to an inadequate level of the internal transfer potential of research. The following summary of the statistics casts some light on the relevance of this particular potential:
- Only 12.1% of the contributors to a study by the European Industrial Research Management Association had successfully applied more than 50% of their results from 'oriented basic research' (EIRMA, 1982, p. 29).
- Only 21.7% of the German respondents and 33.3% of the U.S. respondents to a questionnaire study have applied more than 50% of their successful basic research in products or processes (Eggers, 1997, Tables 15 and C6).
- 33.3% of the German respondents and 50% of the U.S. respondents to the same study have transferred successful basic research results into development projects.

Certainly, one cannot expect 100% of all successfully completed research projects to be transferred at any point in time. Not all results can be used, and transfer activities themselves take time: both effects reduce the percentage to values below the 100% level. The questions here are, whether technology managers are satisfied with the figures shown above, and how their own organization compares with these figures.

It is a tragic experience for researchers to observe that their offerings of valuable results to other groups within the company are passed over unnoticed or cannot be accepted due to a lack of funds, to the surprise that they create or for other reasons (Ruedi, Lawrence, 1995, p. 516). Speiser, director of the IBM research laboratory at Zurich and later in a comparable position at BBC Corp. in Baden, requires research to stretch out into the spheres of development in order to facilitate transfers. If this transfer is inhibited, the management "is forced to liquidate the laboratory. Many such cases are known. They introduce a fateful tragedy into the lives of researchers who started first class work, but terminated it half way through. It is not they who have caused the mishap - the management has failed" (Speiser, 1971, p. 15, translated by author). Certainly, as long as the sufficient conditions for successful transfer on the side of the development groups do not match the necessary conditions, effective transfers will not be achieved.

The internal transfer potential not only connects different stages of a research or development process. Additionally, it concerns the transfer of knowledge within one stage, but between successive periods of time. This is addressed most impressingly by Nonaka and Takeuchi (1995, pp. 62 et seq.) in their discussion of knowledge conversion within a company. They form a distinction between personal, context-specific, and thus difficult to formalize and communicate 'tacit knowledge', and 'explicit knowledge' which is codified and transmittable in formal, systematic language, samples or documents. There are four modes of knowledge conversion: From explicit knowledge to explicit knowledge, which is called 'combination'; from explicit knowledge to tacit knowledge, which is called 'internalization'; from tacit knowledge to explicit knowledge, which is the mode of 'externalization' and finally from tacit knowledge to tacit knowledge, which the authors call 'socialization'. This is described as "a process of sharing experiences and thereby creating tacit knowledge such as shared mental models and technical skills... The key to acquiring tacit knowledge is experience..." (Nonaka, Takeuchi, 1995, pp. 62 et seq.). The examples given involve solving particularly difficult problems. We assume that the transfer from tacit knowledge to tacit knowledge is an important activity in basic research, particularly as technologies that are demonstrable and that need to be documented, for instance, for patenting are not yet available. Thus, this form of socialization can be of particular importance for the advancement of research processes. It poses particular requirements for human resource management in research departments as opposed to other functional areas.

Some companies quite explicitly address research potentials beyond the creative potential. For instance, supplementary to what has been identified above, some Japanese companies have strengthened the training, planning, and administrative capabilities of their R&D departments. Let us look at two examples. Sumitomo Electric Industries has a 'development planning department' that performs decision analysis for the evaluation of R&D projects, an 'administrative services department' that teaches young managers about technology management, and a 'techno-research center' that engages in technological intelligence analyses. Hitachi's 'R&D promotion office' provides technological intelligence analyses, helps in creating intercorporate projects, and supports planning and business assessment tasks.

Two additional examples from British industry may be added. At British Gas plc 11% of the total R&D expenditure is earmarked for 'strategic research'. This includes 'technology foresight, technology acquisition' and 'maintenance of skills and competencies' in addition to those activities that are process or product related. The activities mentioned match with some of the support functions that were identified above. Another example is provided by Rolls Royce Associates, Inc. (Derby), where 'maintenance and development of skills'-contracts with major supporters of this laboratory are formed. These support the identification potential, the creative potential and the absorptive potential of their researchers. However, in none of these cases all of the potentials are used.

The primary or core potentials support what we choose to call secondary or peripheral potentials. Based on the empirical investigations, we identify three secondary or peripheral potentials:

(f) *Image enhancement*. By this we mean an ability to create or to support a corporate image favorable to the competitive position. Companies that support publishing new research results or that demonstrate and advertise their ability to govern high technology in their field may seek image enhancement.

(g) *Human resource attractiveness*. For some researchers, the perceived working conditions and day-to-day problems in industrial production or development departments are not particularly attractive. Industrial research laboratories may offer conditions that are more to their liking, and at the same time personnel who are attracted to the research laboratories then become familiar with the company culture. This may facilitate their transfer to one of the other departments at a later time, a placement that could not have been achieved otherwise. As was observed in one of our interviews: "Research has traditionally been a recruiting point for staff with technical background throughout the company. We used to maintain a flow of 5% to 10% per year into other areas of the business. With cutbacks throughout the company, this flow has almost dried up. This is in my view a serious problem, because research needs a continuing inflow of new blood."

Here we observe that the potential is of substantial importance to research itself, and not only to downstream activities.

(h) *Support*. Research may offer the potential of winning public funding more easily. While this may not have been of great importance in the past, except for those companies that were highly engaged in the defense business (see Figure 7), it may well gain additional importance in the future. National and supranational institutions support cooperative research in the so-called 'pre-competitive' areas. Spreading the risks, saving cost, and developing network abilities could be objectives that need to be met. This is very often achieved by research rather than by development work.

Often, winning public support may be of interest because it excludes competitors from obtaining the same funding. This may be an important argument in a highly competitive environment, even if that public support is merely consumed rather than invested.

Primary potentials and secondary potentials together define research potentials of a company. Figure 8 summarizes the view that is put forth here.

It is evident from this figure that seeking only contributions of research to new product or new process developments is myopic behavior. Research builds more potentials than those. These other potentials need to be recognized and accepted by the top management of a company. For this purpose it is helpful to negotiate a 'master research plan', a 'research charter' or a 'mission statement' for the research group, among its members as well as between its management and the top management of a company. This document should define "a shared understanding of the mission that research is expected to fulfill" (Rosenbloom, Kantrow, 1982, p. 119; Ruedi, Lawrence, 1995, p. 516). It should name the themes or fields that the laboratory ought to cover, and the potentials that it should try to support or to build up. Such a document is the basis for choosing the strategic directions of research and the proposed projects. It is on the basis of this outline that the potentials need to be addressed in planning. In this respect, little help can be found up to the present. An example of a mission statement that was adopted by a major multinational corporation is given in the Appendix.

Having developed the notions of research functions and research potentials we want to show how these can be applied in planning research. This will be done in the following chapters.

Fig. 7: Distribution of sources of R&D funds

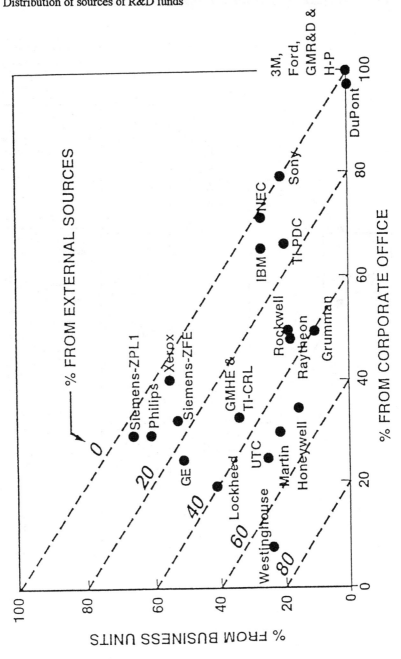

Source: Chester, 1995

Fig. 8: The system of company research potentials

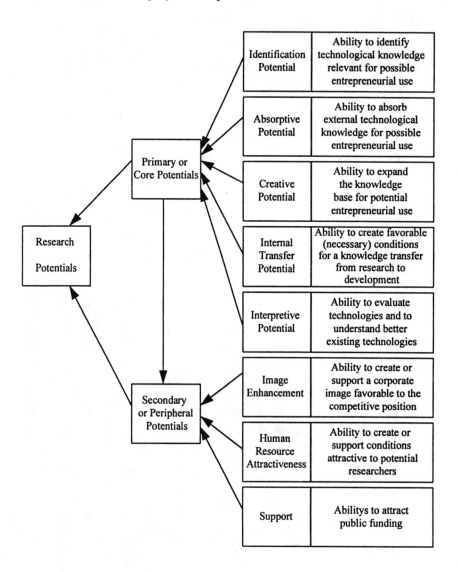

5. Research potentials and project funding decisions

At first, we shall restrict ourselves to the project level of research. Research projects should not be selected arbitrarily. Companies have to find out which projects are more important for their future development and which are of less importance. Takeda, who is responsible for Hitachi's R&D activities, calls research that is performed in this sense 'north star research', contrary to 'blue sky research'. The north star and the blue sky are both far away from the earth, but only the former offers a definite direction. The blue sky would only signal that research goals are far different from where we stand now, but provides itself no specific direction. The same views were expressed by German research managers. One of them says: "We don't do things that live in cloud-cuckoo-land" (Berthold, 1968, pp.- 198, 200). Here again, the idea of choosing a direction for company research is clearly voiced: research needs to be targeted.

While the concept of research potentials may be of interest for its own sake, it can also be used to plan the funding of research in a bottom-up approach. Empirical research indicates that such funding is quite common in some countries. We could identify the funding procedures for central laboratories of 53 large German corporations. These laboratories are most likely engaged in research, although not exclusively so. With respect to the sources of funds, four different funding schemes could be identified:
 (a) Funding from company overhead.
 (b) Funding by taxing business units, not related to the level of support demanded by these units.
 (c) Funding by taxing business units related to the level of their support.
 (d) Funding by selling projects to business units.

From Table 6 we conclude that none of these strategies were applied in a pure sense. In this respect, funding by way of schemes (b) and (d) are of particular interest. The primary funding source is substituted to a substantial degree by other sources. Research draws on 'subsidies' from some kind of overheads in the case of (d), and from project related payments in the case of (b). The highest share of research is reported for strategy (b), followed by strategies (d), (a) and (c). It should be noted that funding by scheme (d) does not contradict performing research at a substantial level.

The funding schemes shown in Table 6 appear to have weaknesses. Funding not related to performance is more frequent than other funding rules. Funding from overhead or taxing business units is correlated with non-favorable perceptions of effectiveness or efficiency of central R&D laboratories. Other funding procedures could be more preferable.

Funding in individual companies may digress from the statistical data given in Table 6. At Daimler Benz, 7.4% of the total R&D budget is earmarked for long-term precautionary research, where equal shares have to come from the holding and the business units. It is known that this 'forces' some business units to spend on this type of research, while their management might prefer to use the funds to boost their short-term earnings. The research plan is discussed during a one-day retreat with the whole board of directors. There are four major sources of funds: (1) funds for long-term projects suggested by central research, discussed with the business units, and with priority determination during the retreat; (2) funds for medium-term projects suggested by business units, discussed with central research, and with priority determination by the research committee; (3) funds for projects entirely at the initiative of business units; (4) contributions to funds from outside sources that need to be integrated with internal funding. This may serve as an example for mixed approaches to the funding problem, but it is still almost entirely project-driven.

Siemens AG recently adopted a funding scheme for central research and development whereby one third comes from corporate overhead, and fifty percent are provided by the divisions. This is tied to the freedom to subcontract their research and development work to outside organizations. The scheme is different from earlier procedures. Then, two thirds of the budget came from corporate overhead. It is observed that the new scheme has not led to a drastic decline in the central laboratories' revenues. However, it initiated a shift towards less traditional fields and towards more software-related research (Wagstyl, 1996).

Table 6: Funding procedures in central R&D laboratories of 53 large German companies (% of funds coming from different sources)

Sources of funds	Group (a)	Group (b)	Group (c)	Group (d)
Company overhead	89.7	1.1	0.0	10.2
Taxing business units, unrelated to level of support	0.6	68.0	0.0	10.6
Taxing business units, related to level of support	0.3	7.8	87.0	10.1
Funding project-wise	3.2	15.1	8.2	62.2
External funds	6.2	8.0	4.8	6.9
Relative frequency of units	30.8	17.3	9.6	42.3

Source: Warschkow, 1993, p. 190.

(a) Functionally-oriented, project-based research funding

We suggest to adopt strictly project-based funding procedures that relate to all potentials that could be generated by the research. This has at least two sorts of benefits. First, it makes the application of the functionally-oriented approach easy. The approach would be spoiled by generous funding from corporate overhead. Second, it increases communication between research and its customers (Warschkow, 1993). Apparently, this can help to grow the internal image of research and its importance for competitiveness, because research is forced to leave the ivory tower and to advertise its strengths. Ultimately, research becomes a profit center or an investment center. The time-horizon for funding decisions should be project-specific and not standardized to conform with the usual one-year budgeting rules. Projects could then range between very long term projects and shorter term projects. This flexibility needs to be stressed. Reportedly, a similar project funding mechanism that was used by Philips but did not have this flexibility had to be abandoned. Furthermore, each project could be funded from more than one source of funds. The shares that the different sources contribute to a project would depend on the project's contributions to the different research potentials of the company.

Such shares are shown in Figure 9 for some artificially created sample projects.

Fig. 9: Example of funding by projects and potentials

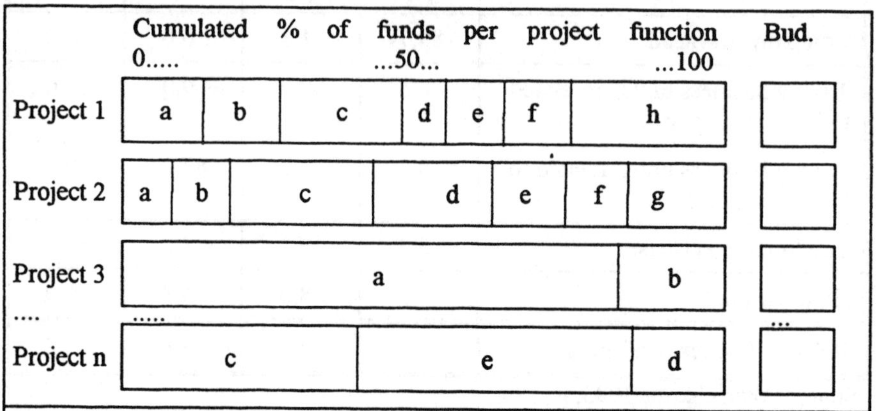

	Cumulated % of funds per project function							Bud.
	0.....			...50...			...100	
Project 1	a	b	c	d	e	f	h	
Project 2	a	b	c		d	e	f	g	
Project 3	a						b	
....	.....							...
Project n	c			e			d	
Bud.: Total project budget; a to h refer to the potentials listed in chapter 4.								

The project objectives, the total project budget and its time frame need to be planned as in all project management approaches. In addition, it is necessary to estimate the shares of funds that must be allocated to the different potentials that could be derived from the project objectives. By considering all research functions and potentials, projects are given a much broader funding base.

The funding is not narrowed down to the view that specific product areas need to be supported by new technologies. Thus, a broadly based view of contributions to competitiveness can be realized.

Let us now consider the sample projects. We begin with project n in Figure 9. We recognize a relatively high potential to expand the knowledge base (c) and to create favorable conditions for transfer (e). A small contribution of the project is expected from understanding better the presently used technics (d). Funding for such a project could most probably be secured from one or more business units.

Project 3 suggests increasing the identification potential (a) and the absorptive potential of research (b). It addresses interests that are most likely not held by the management of existing business units, but rather by the top management of the company. This might be a long-term investigation of some technology that is considered relevant for the future development of the company, even beyond the presently defined technologies of different business units. Top management could fund such a project on a multi-year basis. It is obvious that this type of project cannot hope to attract funds if it is not covered by the charter or the mission statement of the laboratory. This exemplifies the interaction between this document and project selection decisions.

Project 1 addresses almost all potentials, and could be used, in particular, to generate public funding (h). This may be due to the substantial possibilities in generating a strong identification potential and a good absorptive potential. Project 2 appears to be more company specific and product related, and is therefore unable to generate public support. It should attract the interest of at least one of the business units.

Figure 10 presents a functionally-based, project funding form that was developed on the basis of these ideas. In its rows are listed the potential project functions. One should first indicate whether such functions are likely to be supported by the respective project or not. Then, possible sponsors for such functions should be identified in the further columns. After that, funding shares for the respective functions and their possible sponsors can be entered, possibly as the result of negotiations between the project initiator and possible sponsors. The example shown is taken from a larger number of test-runs for the form that were applied within cooperating companies. In some cases, the funding of completed projects could be replicated by the use of the form, while in some other cases the form initiated discussion on possible other sources for project funding then the traditional ones. This suggests that the form may be used not only in the project planning phase, but also as a controlling device in later stages of project development - in this case adding information on the divergence of actual from planned funding data.

Project funding is not made easier by this approach. However, determining project contributions to research potentials helps identify potential project 'customers' at the same time. Research management and researchers would then have to ask themselves what potentials any one project supports and, consequently, which 'customers' should be applied to funding. Funding decisions should not be standardized with respect to the period of time for which funds are allocated. Depending upon the project's characteristics, both long-term funding agreements and short-term funding agreements may well co-exist at the same time. Top management may initiate projects that aim at identifying potentially important technologies by collaborating with universities and scanning the technological environment. It may also provide seed funds that help to integrate new researchers into the company. Contrary to this, business units or their development departments may use the research group like a contract research institution. For example, Shell Internationale Research Maatschapij (SIRM) is considered "as contractor to other (operating) companies"*. The research department could help evaluate presently used technologies or help develop creative ideas for future entrepreneurial use. This could involve projects performed by small groups in short time intervals. Thus, project funding is tied closely to needs of internal customers, and research management would be forced to identify these needs if it wants to sustain or to increase its research capacity.

(b) Organizational Support

The project budgeting procedures outlined above need organizational support. It was already mentioned that our suggestions support a profit center or investment center solution for research. As an example of the relationships between a central research laboratory and the business units of a company, we present a scheme that describes Nokia's solution (Figure 11). The budgeting scheme needs to be completed by a layer of 'informal technological relationships' that is strongly supported by the management.

---

* From an interview, Jan. 29, 1995.

Fig. 10: Project funding form

Potentials/functions of research	Contribution to these potentials?	% of Bgt.	Possible sponsors for the project budget									
			Internal sponsors (Top Mgmt., Business Units, Dev. Labs, Other Central Depts., ...)										External sponsors	
			Corporate		BU		DevLab		Central Lab		....		Government	
			1.000 hfl	% of Bgt.	1.000 hfl	% of Bgt.	1.000 hfl	% of Bgt.	1.000 hfl	% of Bgt.	1.000 hfl	% of Bgt.	1.000 hfl	% of Bgt.
Identification of external technological knowledge	o yes o no	10	314,0	30	523,3	50	104,7	10					104,7	10
Absorption of external technological knowledge	o yes o no	10	314,0	30	523,3	50	104,7	10					104,7	10
Creation of a broader knowledge base (new products/processes)	o yes o no	30					2512,1	80	628,0	20				
Creation of favorable conditions for transfer to development	o yes o no	20			1256,0	60	418,7	20						
Evaluation and better understanding of existing technologies	o yes o no	20					1674,7	80	418,7	20			418,7	20
Improvement of image	o yes o no	10	418,7	40	628,0	60								
Improvement of attractiveness for new personnel	o yes o no	-												
Attraction of outside funding	o yes o no	-												
....	o yes o no													
Sum (Shares in %)	must add up to 100%			10%		28%		46%		10%				6%
Sum (1.000 hfl)		10467	1046,7		2930,8		4814,8		1046,7				628,0	

Fig. 11: Interdependencies between Nokia Central Research (NCR), top management and the business units (BU's).

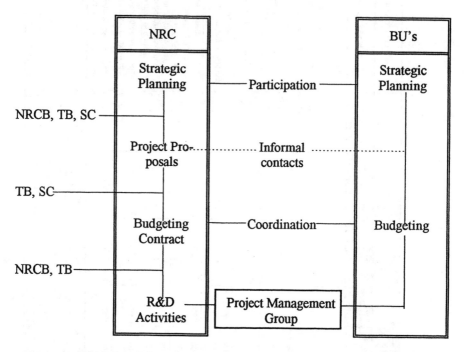

NRCB = Nokia Research Center Board, composed of CEO, top management representatives from business units, top business units' R&D management.
TB = Nokia Technical Board composed of top management representatives from business units, top business units' R&D management.
SC = Technical Area Steering Committee, composed of top and middle business units' R&D management.

If projects of a central laboratory need budgeting support by business units, this support has to be in line with the units' strategy. To avoid costly surprises, strategic intent has to be communicated to and coordinated with the central research laboratory. Thus, a frame for its project proposals is created. Such planning loops help to avoid surprising go-stop-go-decisions that can otherwise lead to ineffective and inefficient research work.

The approach described here does not require the existence of a central research laboratory. For instance, Finmeccanica does not have a central research center for historical reasons. Individual laboratories

are operated as centers of excellence for certain technologies. They differentiate between three types of projects:
- Type A: Preliminary research (long term, highly innovative, high risk)
- Type B: Finalized research (mid term, innovative, medium risk)
- Type C: Development (short term, slightly innovative, low risk).

In 1994, the expenditure allocated to Type A and Type B projects was 32% of that for the Type C projects, or 24% of the total R&D expenditure. The coordination among Type A and Type B projects is achieved by a Management of Technology Group that provides a Technology Matrix (where 860 technologies are related to 140 business units, together with strengths and weaknesses in performing these technologies), Technology Reports (which assess internal know-how against the state of the art in relevant technologies), and a Technology Plan. Managerial processes have been established for coordination among the centers of excellence on the basis of the planning materials (see Figure 12).

Type A and Type B projects may involve outside partners, and they may generate support, such as from the European Union. Again, multi-functional and multi-sponsored projects are possible. However, with the lack of a central research unit it appears that more coordination work becomes necessary.

With the arrival of centers of excellence planning and project budgeting seem to request more organizational provisions than in the case of a central research laboratory. Here, the knowledge on the technology matrix and the technology reports are instrumental in guiding suggestions for research topics to the top management and the business units. Multiple funding sources for a project could still be possible and possible sponsors may be identified by the technology matrix, because it reveals if one technology is of interest to many businesses.

The major point to note here is that there exist interdependencies between the requirements of project budgeting schemes and organizational structures. These should be developed to the best of the company.

Fig. 12: Finmeccanica's process of technology planning

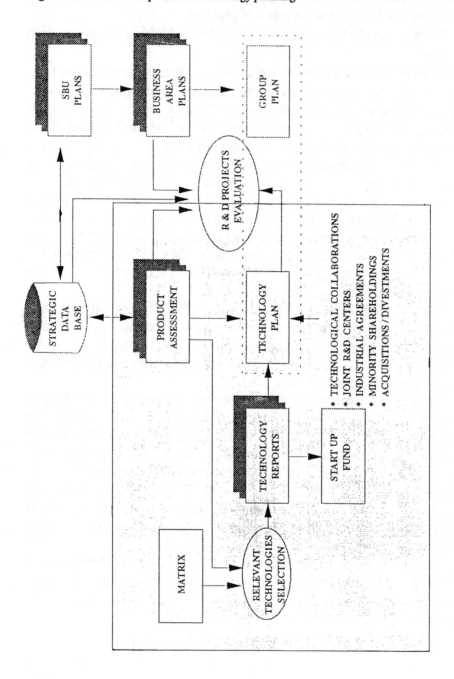

6. On property rights and project potentials

As mentioned with reference to the Finmeccanica organization, projects that build more than one potential could be funded by more than one source. This inevitably leads to the question: who 'owns' the project? A convincing response to this question is that each funding source owns a share of the project equaling its relative contribution to the total project budget, as agreed upon between the laboratory and the sponsors, unless some other sharing rule has been negotiated. Budget overruns or budget savings may be experienced due to the uncertainty that is inherent in research work. In our view, benefits and costs of the uncertainty should be enjoyed or borne by the laboratory.

Should research be mistrusted to overcharge its internal customers this could be checked by auditors. This is standard behavior in publicly funded university research. It could be balanced by allowing for competition. This could be internal competition among research groups in big companies. It could also involve external competition in all companies, with the provision outlined above that at least the identification potential and the absorptive potential need a longer term funding perspective, possibly from a board of directors or a similar top management group.

Regarding the specific character of knowledge, namely the possibility of using it multiple times without additional or marginal cost, a further issue must be solved. There needs to be some regulation of the conditions under which the laboratories or the sponsors could generate additional income from selling or leasing the new knowledge from a research project to third parties, be they internal or external to the company. A solution to this problem may be found by giving the laboratories the option of offering different types of project contracts to their sponsors. If the sponsors reserve for themselves the right to make multiple uses of the project results, the project budget should cover the laboratory costs, a surcharge for consequences of uncertainty, and a possible laboratory profit. If the laboratory reserves the right to make multiple outside uses of its results, it should not try to add a laboratory profit to the other budgetary items, as such profits could be derived from the additional sales of the new knowledge. Laboratory profits could be used - at least in part - to sponsor projects that the laboratory feels are necessary for its own sake.

In this respect it is interesting to return to the case of Shell. It is reported: "SIRM itself has a research budget (internally referred to as General Research) which is more or less equivalent to what elsewhere might be called corporate research. This budget is raised by a levy on the operating companies, the level of which is set by a Committee of Managing Directors. General research is often longer term and more generic. Increasingly in recent years, the business sectors and operating companies have been encouraged to participate in planning the General Research, but the decisions rest with SIRM, again at board level"*.

Our suggestion includes the right of potential sponsors to place projects at external laboratories as well as internal laboratories to balance possible overcharging. The right of the sponsors for utilizing external knowledge suppliers should be buffered by, first, a rule according to which the internal laboratories are given the option of developing and making offers, and second, by general criteria for the selection of competing offers. The latter may be difficult to formulate. While criteria such as cost and duration of a project may be formulated easily, the evaluation of competence and of credibility in solving a given problem can hardly be operationalized. However, the same difficulty arises in the choice of other types of contractual relationships as well. Therefore, it cannot be held, in principle, against the suggestion developed here.

In fact, business units at Daimler Benz intending to sponsor long term research projects may solicit proposals from the outside, but must give their own central laboratory the right of last call. This comes extremely close to what we suggest. Efficiency of research is continuously monitored by collecting data on customer satisfaction, while effectiveness is additionally the concern of audits that even employ outside experts and that are performed as it appears useful.

---

* From an interview, January 29, 1995.

7. Research potentials and relative share of research

The identification potential, the absorptive potential, and the creative potential can all be made immediately relevant for entrepreneurial use. They aid in generating new business. In this section we concentrate solely on these potentials. The question arises of what share of the total research and development expenditures should be earmarked for these potentials. In this sense an optimal minimum share of research is sought, assuming that the other functions play no role in this respect.

In economic terms, the ability to generate new business from the creative use of research can be expressed in a very compact manner by the elasticity of research. An elasticity is the quotient of the relative change of some output measure over the relative change of some input measure. Here, output could be sales or value added, while one of the inputs is research. This is taken as a proxy variable for the knowledge base derived from research. Taking riskiness of research and its long lead time into account, the elasticity can only be determined in a long-term view and at an aggregate level, not for individual projects.

The concept of elasticities is an important one in economics. The reason for this is that companies are managed optimally, when all inputs utilized have the same marginal productivity. This marginal productivity can be derived from the knowledge of the elasticities of each factor that contributes to the generation of outputs. Therefore, elasticities are used as a basic information to determine optimal strategies for pricing, advertising, factor shares in production etc. Managers in these fields are used to this concept. This is not, however, the case for R&D managers. Our interviews showed that managers had to be pressed hard to think in terms of an output elasticity of research, and the estimates that they then produced were given with much reservation.

7.1 Mandatory research

At first we want to consider a situation, in which there can be no generation of output without research. We call this type of research mandatory research. It is an indication of mandatory research, when the question is asked: "Why is it that we can no longer rely solely on engineering applications?" (Ruedi, Lawrence, 1995, p. 512).

To model the situation we assume a generally applicable relationship between inputs and outputs, namely a Cobb-Douglas-type relationship. Output is operationalized by sales. Inputs are all other factors of production, research, and development. Development and all other factors are assumed to effect output in the same period, while research only effects output with a lag of k time periods, for instance k = 10 years. The relationship between inputs and outputs is optimized under the further assumption of a long term equilibrium: we are only interested in an equilibrium distribution of funds among research and development that avoids the pendulum swings identified in chapter 2 as detrimental to company development. A detailed description of this model is presented elsewhere (Brockhoff, 1995 a).

Under these conditions, important results on the optimum share of research (R) over development (D) expenditures can be derived. This share should be

$$\frac{R}{D} = \frac{l}{h} \cdot \frac{1}{(1+i)^k}.$$

where
- l   is the output elasticity of research,
- h   is the output elasticity of development,
- i   is the interest rate, and
- k   is the time lag between research and its use in the generation of output, while no such time lag is assumed for development.

The relative share of research depends on a quotient of elasticities, and a term that takes into account the effects of the time lag between research and its use. Relatively more will be spent on research if its elasticity increases relative to the other elements of the formula. In addition, relatively more will be spent on research, if interest rates or the time lag decline. Both effects work in the same direction and reinforce each other. If interest rates (or the cost of capital) increase, research will come under pressure either to speed up the transfer of its results into new products or new processes, or to reduce its share in the total R&D expenditures.

The relation of research to financial conditions is very interesting, and is by no means only marginal. If the cost of capital rises from 10% to 12.5% and if it takes 10 years from research to market effects, this would result in a 20.1% decrease of the optimal research expenditures. Alternatively, if the time lag could be reduced to just eight years, the original level of research could be maintained.

While managers may not apply our model and the optimization approach in exact terms, they can develop reasoning related to the results derived here. Again, the Alcoa case provides a very telling example: "In the early years of the depression, the executives on the General Committee virtually eliminated new fundamental work..." and even the research director wrote that "in the present state of business it is essential that we concentrate as much as possible of our energy upon lines of research which are likely to be productive of immediate benefits to the Company .... Some Alcoa executives believed that (the laboratory) had no business doing any kind of fundamental research" (Graham, Pruitt, 1990, pp. 211 et seq.). The cost of capital changes during business cycles, and it depends on the level of risk involved in an investment. It is an environmental effect that is not determined solely by the company in question. The quotation illustrates that managers relate environmental effects to research, even though they do not use the model explicitly.

During a business cycle research may be influenced not only by an interest rate effect, but also by a liquidity effect. It is certainly not easy to discriminate between the interest rate effect on research shares and the liquidity effect that may restrain the level of research expenditures. Again, a case can illustrate this point (Erkner, 1996, pp. 64, 73). At Continental AG, a major tire manufacturer, an internal memo addressed to the supervisory board in March, 1970, points at the importance of basic research in chemical physics and chemistry for the overall progress and success of the company. As a consequence of dwindling market shares and sales a completely different approach appeared advisable only about half a year later. Although the strengthening of research and development would have been required its budget was cut by almost 20% over the next two years.

The small basic research group feared for its very existence and it bombarded the directors with desperate memos.

Here, we see clear signs of a liquidity effect that works independently of the interest rate effect.

7.2 Mandatory research and transfer cost

Let us now assume that there is a cost attached to transferring research results into development. This cost is inversely related to the internal transfer potential. Therefore, it may increase with an increasing level of research results to be transferred, and it may decrease with an increasing level of development activities, as this expands the absorptive capacity of development. (This could be expressed as wR/D^f, with positive parameters w and f). The share of research then becomes smaller than shown before:

$$\frac{R}{D} = \frac{1}{h} \cdot \frac{1}{(1+i)^k} \cdot \frac{1}{1+z} = \lambda.$$

Here, z depends on the transfer cost parameters, the elasticities, the interest rate and the time lag*. If transfer costs are observed, z becomes larger than zero. Therefore, relatively less will be spent on research. The whole expression on the right-hand side of R/D is called λ for easier reference in later expositions.

The level of the transfer cost may be influenced by very many variables. It may be seen as a result of interface problems that exist between research and development departments, i.e. a lack of a sufficient internal transfer potential. These problems may arise from 'different orientations and expectations' regarding the roles to be played by the respective departments, which could be further nurtured by lack of communication (Burgelman, Sayles, 1986). Lack of communication may result from too great a distance between the research site and the development site or the location of other in-house customers of research (Allen, Fusfeld, 1976).

In 1966, Deutsche Shell AG closed down its basic research laboratory at Birlinghoven close to Bonn. It was organized as a separate

---

* It is shown elsewhere (Brockhoff, 1995) that $z = (w/D^f)(1+lf/h \cdot (1+i)^k)$.

limited company. Its manager was quoted explaining that it was extremely difficult to coordinate the work of his laboratory with that of the development units of the Royal Dutch Shell group*.

Another remarkable account of these problems is given by Xerox's Palo Alto Research Center (PARC): "Xerox hasn't cashed in on PARC's exciting research on computerized office systems, which was the center's original reason for being" (Uttal, 1983). Many reasons are mentioned for this failure: weak ties to the rest of Xerox, a loose management system that encouraged PARC to overstep its charter, lack of management attention, no channel for marketing products based on the researchers' efforts, and cultural differences magnified by the long distance between the East Coast, where Xerox' headquarters is located, and the West Coast of the U.S. More joint meetings and a joint hiring program that brings new researchers to PARC during their first year before they join other groups have been installed to remedy the situation. The joint hiring policy is exactly identical to what was described above as a secondary potential of research under the title of human resource attractiveness.

BMW AG chose a particular architecture for its 'research and engineering center' (FIZ) in Munich with the explicit aim to eliminate communication hassle by reducing distance between individuals and groups that cooperate on car development.

Similar considerations have to be employed if research results need to be imported from external sources into a company. Again, psychological and physical proximity to the research site appear to reduce transfer cost. This is one of the reasons given by firms for placing research units close to locations where the most interesting university research is performed, and for choosing collaborative forms of research over merely supporting the development of new results by granting money to external research groups. A major reason why NOKIA AB set up part of its central research laboratory in Tampere/Finland was the 'across the street'-proximity to the local university and the possibility of undertaking joint research.

We shall come back to these considerations in chapter 10. Here, it suffices to see that growing transfer cost of new knowledge reduces the relative level of spending for research.

---

* Deutsche Shell AG, Geschäftsbericht 1966, p. 9; Shell schließt ihre Laboratorien in Birlinghoven am 31. Oktober, Siegkreis-Rundschau, August 9, 1966.

7.3 Supportive research

Let us now turn to a different type of research, namely supportive research. Here, some amount of output from product or process improvements may be generated without any research, while more radical, new things can originate from a combination of research, development and other factors of production, as before. This makes it necessary to split total development expenditures into two shares: one to support radical innovations (D - F) and the other to support improvements (F). The optimal share of research expenditure is now determined by

$$\frac{R}{D} = \lambda \cdot \frac{D-F}{D}.$$

λ is the expression that was defined in the preceding chapter (7.2)

For all F > 0 it is obvious that relatively less is spent on supportive research than on mandatory research, because development becomes more prosperous or productive.

Thus, given the above relationships, an important question to ask for long-term budgeting is whether research is mandatory or supportive. The next question should address the level of transfer cost and its manipulability, particularly if the company wants to use external research results. Finally, time lag and cost of capital should be examined, as both of them tend to reinforce each other in influencing the share of research. In any case, there is a need to come up with estimates of output elasticities of research and of development, making these efforts comparable to those of other functional units in the company. This will be covered in (7.4) below.

7.4 Tests of the basic relationships

We have neither knowledge as to the relative importance of mandatory versus supportive research nor as to the level of the transfer costs that may arise. Still, it is interesting to note that the share of R/D should depend linearly on a variable $1/(1+i)^k$ without an intercept and a coefficient that depends on l/h as well as the other parameters that may have to be observed. It would be an important test of our analysis, if such a relationship could be established from empirical data. This will be tried for the German industry.

We would like to determine a positive, unknown parameter b from the equation

$$R_t/D_t = b \cdot (1/(1+ i_{t-m}))^k.$$

The data for the dependent variable are available for the years 1965 through 1991 (Echterhoff-Severitt, 1988 and subsequent volumes). The independent variable is composed of the interest rate and the time-lag k. The interest rate was taken from available data (Dresdner Bank, 1991). A different choice of the interest rate variable can affect the level of b. The interest rate enters the equation with a time-lag of m years. As it is unknown whether management would consider in its planning the interest rate of the previous year (m = 1) or that of two (m = 2) or three (m = 3) years ago, we performed alternative calculations for these observations. Furthermore, we have no general information on the time-lag k. We chose to set k = 1,2, ..., 12 and run alternative calculations. The statistically best estimate of the unknown parameter b should then be selected. The selection criteria are the significance of the parameter's difference from zero, and the coefficient of determination.

It could be established that the relative share of the industrial research budget does in fact correlate very strongly with the level of the interest rate (see Table 7). This is evident from the high T-values as well as the high coefficients of determination. Similar results can also be obtained on an industry level (see the Appendix).

The result is considered very interesting. It has substantial managerial implications, as has already been mentioned.

Assuming that the research budget is planned at least one to two years in advance, and that these plans are based on past observations of the interest rate or the cost of capital, we find that the most plausible estimate of the time-lag of the research effectiveness is between three and ten years*. In fact, as reported from the planning proce-

---

* Alternatively, we have estimated both b and k from nonlinear regression by the Levenberg-Marquardt algorithm (Nash, 1987). The best result by the sum of squares criterion is achieved with interest rates lagged three years. The parameters are very close to the results in Table 5. We get: b=0.1090 (T=5.11); k=10.3923 (T=3.87); sum of squares = 0.000293.

dures observed for the research group within the Shell organization: "The planning process is based on a two years cycle. In year x, the programs for years x+1 and x+2 are defined. In year x+1, high level reviews take place of the results of the previous cycle and of progress to date in year x+1. The programs for year x+2 are then adjusted (in principle only marginally) in the light of the reviews and any intervening changes in the business requirements."* We observe that the estimated research time-lag is very sensitive to the assumption of the time-lag for the interest rate used in the planning procedure. Thus, more information on the actual planning processes are highly desirable.

Table 7: The relationship between the share of industrial research expenditure and the interest rate in Germany, 1965-1991.

Time-lag of the interest rate variable (m) (years)	1	2	3
Time-lag of the research effect (k) (years)	3	6	10
Regression parameter (b)	0.063	0.079	0.106
Standardized regression parameter (beta)	0.988	0.990	0.995
T-value for the regression parameter	21.09	23.11	34.11
Level of significance for the regression parameter	0.0	0.0	0.0
Coefficient of determination (R-square)	0.9759	0.9798	0.9906

If actual industry behavior is close to optimizing behavior, the elasticities of research and of development should be equal to the respective intensities (R&D/sales-ratios). Assuming that approximately 3.1% of sales are spent on R&D in German industry (1991) and that 5.7% of total R&D is spent on research, the ratio of the elasticities l/h should then be 0.057. As only a minority of firms engage in research, the ratio ought to be higher for these firms. This ratio should be compared with the regression parameter b. The empirical data show higher values than 0.057. But, considering the argument that has just been raised, these results are within a plausible

---

* From an interview, January 29, 1995.

range. Because the data used here are rather fuzzy, the empirical 'test' should not be given too much weight.

7.5 Estimates of research elasticities

A major question is whether the elasticities we have used above can be determined in practice. We can see at least three different approaches having their own particular pros and cons:
- Elasticities can be estimated from long term time series data of individual firms (Brockhoff, 1994, pp. 229 et seq.). Practical experience has shown that it may take six weeks to collect the relevant data from company files. The resulting estimates are valid only to the degree that no major changes in technologies occur in the present or future and that no major new markets are opened. These are very heroic assumptions with respect to research.
- Elasticities can be estimated from cross-sectional data separately for research and for development activities (Mansfield, 1980; Link, 1981; Griliches, 1986). To capture the different time-lags between research, development and outputs a panel approach is most suited. However, this involves comparing between competing firms. Inasmuch as research provides singular and exceptional competitive advantages, these can hardly be captured by the approach. Results from the approach can be used as a baseline for subjective evaluations.
- Elasticities can be estimated from perceptional data that can be collected from individual research representatives. In the context of the present research project this was achieved by Bardenhewer (1996 b). His approach will be presented in some detail.

Asking research managers to estimate elasticities meets with surprise and resistance. They are not used to the concept of elasticities and they lack a baseline for their estimates. This may be a consequence of traditional funding rules for research, whereby allocations of company overhead to research were not based on explicit considerations of elasticities and marginal products of research expenditures. Therefore, a multi-stage interview procedure, the so-called Delphi-approach, appears to be particularly well suited for the organization of the interviews. After an initial round of questioning, the median of the responses is fed back to the participants along with commentaries to explain extremely low or high estimates. The respondents may then revise their responses in a second round of questioning. This procedure allows participants to learn by familiarizing themselves with the concept of elasticities and by comparing their views with those of other respondents in the same industry. It is well known that this procedure leads to a variance reduction in the responses as well as a possible shift of the median. Experimental research indicates that this shift of the median may or may not approach a true value for the responses, which is one of the issues criticized in Delphi-styled interviewing techniques (Linstone, Turoff, 1975; Brockhoff, 1979). In spite of this the Delphi-method is widely used for long term forecasts of technological developments*.

For the present study 39 technology managers in Japanese and European companies from the electrical and electronic industry were invited to participate. We mention the industry here, as we assume that the results will be industry specific. The major task is not to demonstrate differences by industry, but to demonstrate whether the Delphi-method is a feasible approach for the estimation of elasticities. In the first round of personal interviews, 30 managers agreed to make estimates. The results were fed back in numerical and in graphical form on computer disc, which was also used to collect the estimates in the second round. In this round, 17 managers participated. It is not unusual in Delphi-styled interviews that the number of participants drops from one round to the other. This could be another reason for concern if the dropout is associated with some sort of bias. It was impossible to check this in the present study.

Respondents were asked to give three estimates at six levels each. The estimates concerned the relative change of development expenditure as a consequence of a relative change of research expenditure, the relative change of sales as a consequence of a relative change of research expenditure or - in addition - as the consequence of a relative change of development expenditure. In all cases the respondents were asked to think of lasting, long term effects, and to assume that all other things remained unchanged. The six levels to be considered were reductions by 5%, 10% or 20% of the respective research or

---

* See for instance the German-Japanese comparison of long term technological trends: Bundesministerium für Forschung und Technologie, 1993.

development expenditure, and - alternatively - increases in the respective funds by the same relative levels.

Results are summarized in Figure 13. Technology managers in the electrical and electronics industry find that research is substitutive to development, all other things remaining equal. This can be read from the upper part of Figure 13. If research budgets are reduced by a particular percentage share, development budgets need to be increased. The reverse is true if research budgets are increased. The research elasticity of development is negative.

Customarily, the share of development is much higher than the share of research (see also Figures 1 and 2). The numerical values indicate that a reduction of research by 10% from the present budget level of 5% or 10%, respectively, would necessitate an increase of the development budget by 1.9% or 1.8%, respectively.

The total budget for research and development would increase by 1.4% or 0.8%, respectively. Research cuts in reality are not likely to be compensated in this way. This could lead to long term sales declines, as the 'all other things remaining equal'-assumption is no longer met.

There has long been speculation that the research elasticity of development could be negative, the major argument being that "the results of research can be used to predict the results of trying one or another alternative solution to a practical problem" (Nelson, 1959, p. 299). This means that development can more securely start from an advanced knowledge level that eliminates trial and error. This should be true if technological knowledge is cumulative rather than empirical. It is argued that in the electrical and electronics industry technological knowledge tends to be cumulative, while it is empirical in the chemical and pharmaceutical industries. On this background, it comes as no surprise that the results reported here are hardly matched for a sample of firms including those from industries where technological knowledge grows empirically.

Let us now look at the research elasticity of sales which is shown in the middle of figure 13. This elasticity is positive, i.e. future sales will increase if research expenditures are increased, and they will decrease as research expenditures are reduced. Comparing the middle part of the figure with its lower part, we see that the impact of a certain relative budget change in development is assumed to be higher than in research. While the research elasticity of sales is about 0.5, the development elasticity of sales is almost twice as high (close to 1.0). Again, considering the much lower level of research budgets than development budgets, the cost of achieving a 10% increase in sales is much lower when attempted by increasing research than by increasing development. Starting with a 5% share of research, the total budget would have to be increased by 1%-point in the former case, but by 9.5%-points in the latter case.

The results demonstrate that it is possible to use the concept of elasticities at an aggregate level and for strategic decisions, such as the determination of a long-term budgeting strategy for research and development.

The formula for the optimal share of research over development in chapter 7 (7.1) can be reformulated to yield the optimal share of research over total research and development expenditures. If we use the data from figure 13, assuming a time lag of 10 years for research and an interest rate of 20% due to the relatively high risk involved, this share can be calculated at 7.5%. If the time lag or the interest rate are lower, the optimum share will reach a higher level. Compared with the formulae in the sections 7.2 and 7.3, above, we know that the formula in chapter 7.1 provides an upper limit for the optimum relative research expenditure. Given our assumptions, the result is in an order of magnitude that conforms to the observations of some firms. It can be considered as a proof of reliability of the responses. At least, it gives no reason to reduce research expenditures below the present level.

Still, the results reported here should not be seen as representative of the whole industry. Mansfield's econometric analyses result in value added elasticities of research that are considerably higher than the value added elasticities of development (Mansfield, 1980, p. 865, Table I). Mansfield's study covers many industries with different types of technological process. It is in contrast to the industry-specific results obtained by Bardenhewer. It is obvious that more experience and more study is needed in order to come up with truly reliable elasticity estimates, whether with respect to sales or value added.

The importance of valid elasticity estimates can hardly be overrated. Results achieved so far are based on the reactions of a small convenience sample of technology managers in the electronics industry. They serve to demonstrate the feasibility of the approach. Within

Fig. 13: Relative impacts of relative budget changes (Elasticities)

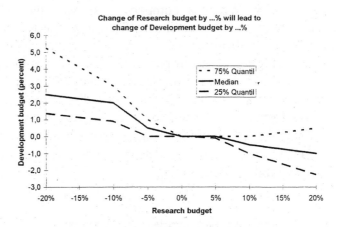

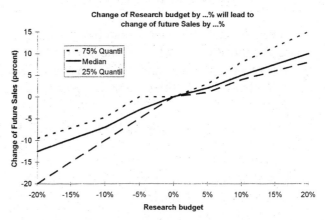

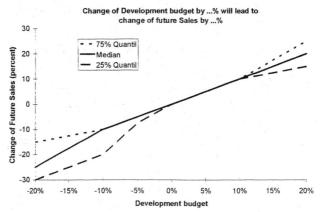

Source: Bardenhewer, 1996 (b)

an industry, a representative sample of respondents would contribute to validity. Within a company, responses by technology managers may be biased in their favor. Therefore, these responses could fruitfully be confronted with those from other departments, and from marketing managers in particular. Differences in the estimates could be used to initiate a discussion that would hopefully lead to a convergent and valid result. Again, such a discussion could be given a specific format. For instance, it could use a Delphi approach if exchanges of results in later stages are accompanied by a collection of reasons for divergencies that may then influence the estimates in further rounds of questioning.

7.6 Limits to research expenditure

The discussion of relative research expenditure or shares of research raises the question of whether the share of research expenditure with respect to sales has any limits. The answer is definitely 'yes'. This will become clear if an optimal level of research expenditures can be determined.

The models of profit-maximizing firms that are used to derive relative shares of research expenditures are based on the assumption that the sales elasticities of research do not exceed the level of one. This implies that increasing research expenditures continue to increase sales, albeit at a diminishing rate. The marginal contribution of research to sales decreases with increasing research expenditure. Considering the partial effect alone, there exists an optimal spending level that is determined by the maximum difference between the linear function describing the increasing research budget and the concave function describing the sales that result from research expenditures.

We shall briefly illustrate these developments. Let us assume that profits (P) are derived from research (R) and an aggregate of many other factors (Z) that have a sales elasticity of $0 < g < 1$. Furthermore, research is mandatory. We would then have the following profit function:

$$P = Z^g R^l - R - Z.$$

As before, l is the sales elasticity of research. By taking the first derivative and setting this equal to zero, we find - after a few rearrangements - the optimal level of research R^*:

$$R^* = \left(l_{Z^g}\right)^{1/(1-l)}.$$

Let us assume that $Z = 100$ money units of inputs, and $g = l = 0.5$. We then find $R^* = 2.2$ money units of research. Assuming returns on sales of 4% we find that optimal research should be about 2% of sales. This is only meant to demonstrate limits to research, and not to serve as a practical calculation. Interestingly, if g were larger this would also lead to an increase of research. The reason for this is that the other factors are used more effectively, and research can therefore be put to a better use as well. The numbers are arbitrary, and are only meant to demonstrate the verbal arguments put forth above.

Another important argument that leads to an upper limit of the research budget concerns the users of research results. If we concentrate on internal users, these 'downstream' departments have to build their potentials to make good use of the research results. Their capacity to do so is limited by the results of their own optimization considerations. If these potentials are lacking, research results lay idle within the firm. The results may be transferred to other firms, either voluntarily and perhaps generating some revenue or involuntarily. The latter may be a direct consequence of the absent potentials in downstream departments. Research personnel who find that their contributions are not appreciated within their own organization may become more inclined to seek employment with a firm which knows better how to use their ideas. This may accelerate the process of losing competitiveness.

This abstract reasoning is further supported by more realistic simulations of company growth models that include competition explicitly. One model, in particular, includes research, development and design expenditures (Weitzel, 1996). In the sense of the preceding models, research is assumed to be supplementary. Holding expenditure levels constant for one competing firm and increasing both the research and development expenditures of a second firm beyond the first firm's levels yields some very interesting results. First, maximum profit levels cannot be achieved by increasing only research or only development. Second, to reach optimal profit levels the firm needs to spend 30% to 75% more on development and 5% to 30% more on research. If more is spent, the firm may enjoy higher sales, but not higher profit levels. If still more is spent, the firm may go bankrupt because it cannot harvest the fruits that it has planted by its research and development spending due to a lack of liquidity.

The results presented here should not be taken as numerical recommendations for the determination of actual budgeting decisions. However, it is possible to formulate similar simulation models on the basis of actual firm data. While the data need to be based on past experience or on subjective estimates, it is not possible to fine-tune the budgets. But experiences with a simulation model that was formulated for a particular firm based on a simplified model of company development (Brockhoff, 1989; Bender, 1997) indicate that companies benefit from knowing whether much higher investments in research or development promise to be profitable or whether they approach the upper limit of further profitable investments in these areas. Technology managers are well advised to consider the use of such models, or special models incorporating specifics of their industry, when making their budgeting decisions. If they do so, they will discover again that the concept of elasticities (or another, model-specific measure of research input-output coefficients) is of great value.

8. On sufficient conditions for research success

Sufficient conditions for research success result from securing effectiveness of research. This should be assured by effective project selection, provided a link can be established between research and the future applications of its results in new products or new processes. As can be seen in Table 5, not all fields of research are of identical relevance to industry. While we have already dealt with literature indicating a relationship between basic industrial research and productivity increases, we shall here use a different line of approach.

It can be demonstrated that the productivity and sales growth of firms is related to the earlier patenting activity of the same firms if 'important' patents are selected (Ernst, 1995). Importance is defined by applications at a foreign patent office or by the number of citations that a patent receives (Narin, Noma, Perry, 1987; Ernst, 1995; Ernst, 1996). For a sample of 26 firms in the electrical and electronics industry it was studied whether research had any correlations with the output of important patents (Ernst, 1997). All results were derived from cross-sectional analyses. The following main results are achieved. Patent applications per research and development 'dollar' over the five year period from 1990 to 1994 are generally unaffected by the share of research expenditures on total research and development expenditures. However, the share of patents granted on all patent applications increases with a higher share of research expenditures. Also, relatively more important patents - as judged by the relative number of citations they receive or a larger share of U.S. patents granted - are granted to firms that allocate higher shares of their research and development expenditures to basic research. Even more interestingly, no comparable correlations could be established with respect to development expenditures. Thus, for one industry, at least, basic research increases potentially valuable output of new technologies.

Tapping the science base for technology advancements can have different importance in different industries. This can be measured by studying the recency and the frequency of literature citations in patent documents. Certainly, classification of the type of literature from which the citations are taken is mandatory in order to avoid using engineering or applied technology literature (Collins, Wyatt, 1988, p.66). Preferably, only literature that reports on basic research results should be used. Fortunately, there exist lists of journals that publish such results. In comparing patents in the fields of pharmaceuticals with those on organic chemical compounds, it is noticeable that the latter use less recent literature; therefore, this field is considered to be 'less high-tech', since it uses a less recent science base (Van Vianen, Moed, Van Raan, 1990, p. 67). The frequency of citations can be used to construct a 'Science Intensity Index' (Grupp, Schmoch, 1992). As shown by Grupp and Schmoch (1992), this index indicates very strong science linkages in the fields of biotechnology, pharmacy, semiconductor technology, organic chemistry and food technology. The lowest linkages are found for home-building and construction technology, machinery elements, transportation, space exploration, medical technology, motors and for turbine technology. Some of these results may come as a great surprise. In this respect, such an index is extremely helpful as a first indication of whether or not substantial contributions towards technology development can be expected from industrial basic research, and consequently whether to invest in basic research if such contributions are of primary importance.

Within a particular field of technology, the selection of preferred projects could be supported by application of the funding scheme that was discussed above and also by further reasoning that would ideally have to be integrated with the strategic planning of a firm. In addition, organizational arrangements can support effectiveness and a strong identification potential. In empirical studies, researchers from large German companies have related two other variables with ineffectiveness (Brockhoff, 1990). The most important variable is a propensity towards overperfection. While this may reflect a specific German engineering tradition or an element of culture, researchers blame themselves for not having enough knowledge of market requirements to guide them in their research efforts. As exact market requirements cannot be outlined to guide research efforts, researchers seem to choose an overperfection strategy as a safeguard against changing or unforeseen requirements. This should not be driven to extremes. Intel's approach, mentioned in Chapter 1, tries to establish a counterweight to overperfection in a rigorous way. The second variable is a propensity for bootlegging. This may have something to do with the first variable, as the clandestine continuation of research projects beyond their official termination may serve to satisfy an engineer's or scientist's wish to perfect or even to overperfect a project outcome. While cases abound in which such projects have led to great success, one should not plan to have these projects and bet one's future on them. It is more advantageous to investigate the reasons for bootlegging, and to analyze whether these could be eliminated.

A further consideration in securing sufficient conditions for a firm's research is the choice of direction. This has been addressed above in reference to the distinction between blue-sky research and north star research. From an analysis of companies which use technology successfully for building competitive advantages, Morone (1993, p. 17) concludes that their management "defines the domain in which their businesses compete. In every case, a strategic focus has been consistently articulated and applied for decades". The history of Roche AG, the Swiss pharmaceutical group, provides one of the many examples to illustrate the point. After an era of a less focused and much self-centered research the company performed a substantial turn from 1986 onwards. Among other things it concentrated its research on selected indications, and within these fields on particular illnesses. J. Drews, the research director, developed a phase model that indicated how many ideas and new chemical entities had to be considered to achieve the admission of at least one new drug per year (Peyer-Roche, 1996, pp. 346 et seq.).

The generalized idea is that a business can be developed successfully on the basis of core technological competencies. According to Hamel and Prahalad, core technological competencies offer
− potential access to many markets
− possibilities for finding unanticipated products
− significant contributions towards meeting customer needs
− protection from competitive pressure, as core technological competencies are difficult to imitate (Prahalad, Hamel, 1990).

The question to be answered is: "Does this project offer the prospect of ... acquiring new knowledge or skills, which would contribute directly to any of our core competencies?" (Coombs, 1996, p. 352).

These competencies should contribute to a vision that explains the indispensability of a company's business. In earlier years supporting strategic intent of a company was a goal similar to what today is called supporting core competencies.

Unfortunately, these benefits of core technological competencies cannot be identified in a mechanistic or automated way. The difficult task of planning for core competencies is reflected in one company's observation that strategic initiatives of business units may not match with its core competencies. It would have to decide whether it should develop such competencies or whether to strengthen existing abilities (Gwynne, 1996, p. 43). No specific criteria are given to take this decision. Furthermore, companies tend to develop 'core rigidities' (Leonard-Barton, 1995, pp. 45 et seq.) or inflexibilities, particularly if they stick to their core competencies for a very long time. This can lead to supporting old technologies in face of technological discontinuities. This has been called the 'sailing ship effect', named for the observation that many sailing ship producers increased efforts to develop their technological competencies when faced with the advent of steamship technology. Similar observations have been made regarding icebox producers, when research had provided possibilities for efficient refrigeration, etc. Furthermore, if core competencies are not rooted in particular technologies, their existence may limit the search for new technological knowledge.

The desirability of supporting core competencies requires that organizational precautions be taken to assure conforming project selection decisions. This should not lead to a committee infrastructure which suppresses innovative solutions and lets only compromises emerge. At Alcoa, researchers demanded at some time to be granted relief from continual goading by innumerable coordinators and expediters, as well as from frequent subcommittee status reports: to reexamine subcommittee structures and responsibilities, minimize duplication of reporting, and thus hoped to facilitate more forward-looking research (Graham, Pruitt, 1990, p. 397). Core competencies have been described as "bodies of technological expertise (both product and process) and the organisational capacity to deploy that expertise effectively" (Coombs, 1996, p. 346). One of the major organizational consequences drawn from an analysis of British research and development departments is to avoid over-decentralization of laboratories. Otherwise, research may be 'damaged' by short-term pressures from the managers of individual business units which may use only a fraction of a particular core competency as each such competency could support products in more than one business unit. This is just another argument for research managers to look around for a multitude of project sponsors.

The sketch of the difficulties in defining sufficient conditions for research success is meant only as a reminder. We want to concentrate on necessary conditions for research success, as these relate directly to the potentials that were identified above. Both, necessary and sufficient conditions need to be met to ensure success. Considering this as a process of management, the view presented in Figure 14 serves as a summary thereof.

Fig. 14: A sketch of the process for securing sufficient conditions for research success

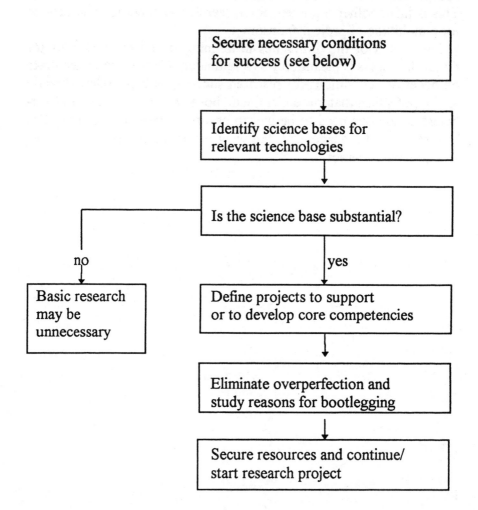

9. Primary research potentials as a necessary condition for research success

9.1 A taxonomic approach

Having identified primary research potentials it is important to ask which of these are necessary conditions for securing company success, and in what combination they should appear. Discussion of necessary conditions for research success does not make much sense if a company only occasionally engages in research: it would then be unlikely that technology managers are cognizant of the need to develop a special research management approach. Supporting this assumption is an observation by Eggers (1997, p. 9). He was unable to discover relationships among variables that explain research functions, communication activities, research success, etc. for a sample including firms with relatively low and relatively high research expenditures. Once he concentrated on those firms having relatively high research expenditures, he was able to support a large number of plausible hypotheses. Although the high spenders had not all adopted identical behaviors, certain common traits could be established. This was not observed for the 40% low spenders in the sample.

Concentrating on the necessary conditions, it becomes obvious that these cannot be defined for research alone. The recipients of the new knowledge, the divisions or departments that are organized further 'downstream' in the flow of the new knowledge, need to be receptive to it. That is, they need to build up their own potentials. This is nicely illustrated by John Armstrong of IBM, who asserts that "we ... all agree that success in R&D is not enough to guarantee the success of individual firms. This remark is an important part of the context for any discussion of the future of industrial research. It reminds us that we must always keep in mind the many factors other than R&D that are necessary for success and that, if done in a first-class way along with effective R&D, will be sufficient to assure success" (Armstrong, 1996, p. 151). Here, we shall address the marketing potential (MA) as an example of other 'downstream' potentials. Whether this potential is created exclusively within a firm or by different forms of alliances, cooperations, etc. should not be of concern in this context.

Furthermore, the relative technological position of a firm vis-à-vis its competitors, established by means of past investments in research and development or by past successes in identifying and transferring new technological knowledge into the firm and keeping it accessible, contributes to understanding necessary conditions for research success. To illustrate only a single point: if the relative technological position (TP) is strong, the firm does not suffer too much from a weak identification potential and/or a weak absorptive potential in the present period of time.

To somewhat reduce the burden of further analyses, we merge the creative potential and the interpretive potential into one, namely an inventive potential (IP). The other primary potentials are used as defined above. We thus arrive at six elements that influence necessary conditions for research and business success:

(1) The relative technological position (TP)
(2) The identification potential (ID)
(3) The absorptive potential (AP)
(4) The inventive potential (IP)
(5) The internal transfer potential (IT)
(6) The marketing potential (MA)

In order to further restrict our discussion it is helpful to assume that each of these elements occurs only in one of two possible states: it is either present (1) or it is absent (0). In reality, we may observe varying degrees of presence of each one of the elements. To illustrate: if one is aware that the technologies being employed have a very weak science base, then a low identification potential is sufficient; if there are indications of a strong science base, this needs to be reflected in a strong identification potential. This idea is taken up in more general terms in section 9.4, below. Presence or absence of any of the elements is considered as a characteristic of a laboratory, not of an individual. We do not believe that one person could represent all potentials at the necessary level. Rather, we assume that a laboratory can obtain the desired levels of potentials by carefully observing among its personnel individual strengths to develop particular potentials. The 'portfolio' of potentials is built from the 'portfolio' of employees.

Furthermore, it makes good sense to assume that the six elements will mainly interact in a multiplicative manner to determine the necessary condition for success or failure. If this is not the case, the absence of any one of the elements could be compensated by the presence of some other element. This is highly unlikely, at least for the present. As we will see later, there are cases in which some kind of compensation may be expected for a future period or a non-innovative approach to management.

Having six elements that can take on two stages each, we arrive at 64 different combinations of these elements. Each combination characterizes a possible research outcome, as well as the strengths and weaknesses that may lead to it. Plausibility allows us to eliminate some of the 64 combinations right away, as they produce non-compensatory failure. These include all of the 32 combinations in which the marketing potential (MA) is missing. Here, research may or may not generate fruitful ideas which are not picked up and transformed to new business. However, we are not concerned in this text with discussing growth strategies for other downstream potentials, which are missing. The combinations with the laboratory environment are represented by the marketing potential. Of the remaining cases, all having only one of the potentials are excluded, as this potential cannot compensate for the lack of all the others. A further combination needs to be excluded, namely the one with a marketing potential and just one of the other potentials. This combination could lead to short-term success selling a gimmick that would sooner or later be discovered as such. Of a further group of 16 combinations all representing weak relative technological positions, 12 of these - having neither the inventive potential nor the internal transfer potential - should not be considered. This is for similar reasons as before. The remaining 19 cases are combinations that are not quite as obvious; they are more interesting, and should therefore be examined in more detail. To facilitate the discussion, these cases are summarized in tables, preceding each of their short textual presentations. We start with Table 8.

Table 8: A characterization of necessary conditions for research success with low relative technological position

Case	TP	ID	AP	IP	IT	MA	Result and characterization
1	0	0	0	1	1	1	? : Autistic researchers who hold promise for the future
2	0	0	1	1	1	1	? : Researchers, deaf to the external, with promise
3	0	1	0	1	1	1	? : Researchers, mute to the external, with promise
4	0	1	1	1	1	1	+ : A substantiated promise for the future

In cases 1 through 4, research offers a strong inventive potential that can be communicated to downstream departments. These, in turn, have the potentials to use this knowledge. However, the present relative technological position is weak. There are no past successes on which to build the future competitiveness. Thus, the company can only rely on the researchers' promises for the future. These promises appear to be substantiated in case 4, in which research is aware of relevant external knowledge and could also make use of it. In the first three cases, research is either unable to identify what happens outside, unable to transfer what it observes, or both. This weakens the promise that its work may offer. The question marks signal that the remaining weaknesses could spoil the internal efforts, particularly if external research results are relevant to the firm and because the firm cannot rely on a strong technological position from past efforts. For the same reason, speeding up the internal efforts is of great importance.

In the remaining cases, we assume there to be a strong relative technological position that was established in the past. This can be used in the future. Thus, even the weak performance of research at the present may not have immediate effects. These effects could manifest themselves later. The following cases differ substantially from the first four cases in this respect.

The next cases can be found in Table 9.

Table 9: A characterization of necessary conditions for research success with high relative technological position (1)

Case	TP	ID	AP	IP	IT	MA	Result and characterization
5	1	0	0	0	0	1	- : The unfruitful hermits
6	1	0	0	0	1	1	- : The trivia communicators
7	1	0	0	1	0	1	? : The hermits who have to be visited
8	1	0	0	1	1	1	+ : Successful research, with the problem of autism

Case 5 describes a laboratory of <u>unfruitful hermits</u>. The group is separated from its internal and its external environment, and because of this or for other reasons the laboratory is unable to develop inventive potentials. For the time being, the company may rely on the technological position it has acquired in the past. Pretty soon, the unsuccessful situation of the unfruitful hermits may be uncovered.

The <u>trivia communicators</u> of case 6 face a different obstacle. They are ignorant about the external technologies, they lack inventive potential, and their internal transfer potential consequently tends to be used to communicate either present work that is non-competitive or reports on past successes. In the long run, this type of a research group may be seen as being even worse than the hermits. It appears to be trivial.

In case 7 research acts like very wise, but lonely <u>hermits</u>, who continue to build a strong inventive potential. The downstream departments would have the capabilities to make use of their research. However, if these departments do not come to visit the hermits and question them, they will not learn about their research results. If visitors come to see the hermits, there is a great chance that they will benefit from such a visit if they are equipped with the potential to transfer the knowledge. The lack of identification potentials and absorptive potentials is no problem for the hermits, if their own potentials are more developed than those of the outside world.

Empirical research shows that these hermits report an extremely low success rate of projects that get to be used by downstream departments, and that they receive only low credibility evaluations from within the company and from outsiders in comparable fields of research (Bardenhewer, 1996). It is disturbing that in the same source it was found that 26% of all laboratories studied exhibited characteristics of hermits.

In practice, then, the question-mark describing the expected success of this type of laboratory or group may actually have to be substituted in many cases by an indicator of failure (-). It appears to be difficult for 'normal' people to communicate with 'hermits', as one does not know how to approach them and as they prefer seclusion.

Case 8 could well be a successful research group, although it suffers from autism. As in the case of the hermits and contrary to case 1, this may not be detrimental, as their past and present potentials shape the competitive environment. Certainly, this position is vulnerable if valuable external knowledge is generated due to the autism from which the group suffers.

A bad constitution of company research is exhibited in case 9, which brings us to Table 10. The stars of the past, with no potential for the identification of relevant external knowledge but with high absorptive potential, may have to take what they can get. Thus, they seek inspiration but they are not selective. Like sponges, these stars of the past absorb what they learn and need to be pressed to reveal some of it. The only way they protect the company, one might say cynically, is by means of their inability to communicate internally. The company cannot rely on the research department. In principle, their situation is no different from the case in which only TP and MA are present, while all the other potentials are missing (case 5).

The activists or the chatterboxes of case 10 are of little help, either. If they use their absorptive and internal transfer potentials without good reason, as they can neither draw on an identification potential nor on an inventive potential, they may appear to be 'wise guys': they will talk a lot, but with little relevance or substance, using the strong technological position their predecessors developed in the past. If the real value of their contributions is discovered, the situation may be remedied. If the truth is not discovered, other departments may be misguided.

Table 10: A characterization of necessary conditions for research success with high relative technological position (2)

Case	TP	ID	AP	IP	IT	MA	Result and characterization
9	1	0	1	0	0	1	− : Stars of the past, mostly sticking amongst themselves
10	1	0	1	0	1	1	? : The activists with a history
11	1	0	1	1	0	1	− : The hermits who relate to whatever comes along
12	1	0	1	1	1	1	+ : Successful researchers with no external orientation

The <u>wise hermits</u> of case 11 use absorptive potentials as do the stars of the past that were characterized in case 9. As do these stars, they accept for absorption whatever they come across. This may disturb the evaluation of their own research results. This disturbance may be among the reasons behind a reluctance to invest in internal transfer potentials, such that few of their results are transferred internally.

<u>Successful researchers with no external orientation</u> are characterized by case 12. Their absorptive potential appears to be wasted, as they have no identification potential. For the time being, the firm can profit from the inventive ideas as well as from the past success. They are deaf or blind to the danger of an erosion of their present position by successfully competing technologies.

Table 11 summarizes the next four cases. We call the laboratories that could be characterized best by the case 13 <u>detectives without influence</u>. They identify relevant knowledge, but they have neither the potentials to absorb it nor to transfer it internally. If approached by their internal 'customers', they may explain the world to them but have little to offer of themselves. In a sense, these detectives seek inspiration as their neighbors of case 9, but they cannot make use of anything they may find.

Case 14 refers to the <u>humiliating researchers</u>. They correctly identify what fits the firm and what they see outside, but they lack potentials to absorb this as well as potentials for substantial contributions to invention. Given their internal transfer potential, this may become clear. Such a situation could have arisen from downscaling a former successful research activity or through unwise staffing decisions. What remains is to communicate external research findings to the downstream departments.

Table 11: A characterization of necessary conditions for research success with high relative technological position (3)

Case	TP	ID	AP	IP	IT	MA	Result and characterization
13	1	1	0	0	0	1	- : Detectives without influence
14	1	1	0	0	1	1	- : Humiliating researchers
15	1	1	0	1	0	1	- : Beavers behind their knowledge dams
16	1	1	0	1	1	1	+ : Successful research, with a not-invented-here syndrome

The non-communicative know-it-alls in case 15 behave like <u>beavers</u>. They very artistically build up knowledge and retain it within their own spheres, preventing it from flowing freely to the downstream departments, quite as water held back behind a beaver's dam. Visiting these laboratories can be more rewarding than visiting hermits. The beavers also know their environment.

We meet <u>successful researchers with a 'not-invented-here'-syndrome</u> in those laboratories characterized by case 16. Every potential is present, except for the potential to absorb external knowledge.

Let us now turn to the final cases, in Table 12. <u>Former stars that are externally dependent</u> are those of case 17. If asked, these researchers can interpret the world outside and even use its new knowledge. On their own, however, they are not very inventive and unable to transfer knowledge internally. They may be found 'playing with the gang' for some time, but if the peers in their field find that they have no entry ticket to the games played, they will soon find themselves excluded from the playground. This will complicate their approaches of knowledge transfer.

Table 12: A characterization of necessary conditions for research success with high relative technological position (4)

Case	TP	ID	AP	IP	IT	MA	Result and characterization
17	1	1	1	0	0	1	– : Former stars, leading to external dependency
18	1	1	1	0	1	1	? : Import dependent research
19	1	1	1	1	0	1	– : 'We-know-it-all'-black holes
20	1	1	1	1	1	1	+ : All-time stars of research

Case 18 represents a sad situation from the point of view of the laboratory which depends strongly on imports of external knowledge. This may not be disadvantageous to a firm that has to choose to exploit an imitation strategy. Lacking inventive potential, but excelling at communicating and identifying new knowledge outside, research may build up on the strong relative technological position and integrate external knowledge to maintain that position. One problem, however, might be that others would do the same. Our firm thus would be well advised to act very quickly so as to stay ahead of others who may use the same knowledge base.

The next case (19) may well be considered to be extremely unsatisfactory. When approached by other departments, these laboratories 'know-it-all' but lack the potential to be active transmitters of their formidable knowledge to the other departments within the firm. Like black holes, they absorb whatever knowledge comes within their reach and actively adds to their own understanding. We suspect that some of the prime industrial producers of new knowledge - set far apart from other departments but in close connections to major research universities in their field of interest - could be classified here.

Finally, case 20 describes the perfect situation: the all-time stars laboratory. It is most likely that every laboratory wants to be in this category. And very probably also, each will discover certain weaknesses putting it into another one. Two questions thus arise: how can missing potentials be identified, and how can they be built up? We shall now turn to these questions.

9.2 Diagnosis of trouble

Missing potentials can be diagnosed if managers are willing to analyze their own situation in an unbiased way. The identification of the potentials is of great help in the first place, as it is much easier to search for indicators of possible trouble if one can relate these indicators to specific potentials. Nursing a general feeling that research does not contribute enough certainly is of much less help, as this attitude could easily lead to throwing out the baby with the bathtub.

To identify weaknesses with respect to the relative technological position of the firm (TP) one may use perceptional measures and objective criteria. The following criteria serve as examples:

- Technology management may evaluate whether the firm is trailing behind its competitors in its technological achievements. In a German study it was found that this relates to over-perfectionism and to delayed decisions on project starts (Brockhoff, 1990, pp. 47 et seq.). A lack of inventive potentials could be another reason.
- If research leads to patents, patent citation analyses could demonstrate whether a given firm more frequently cites patents of its competitors or vice versa (Narin, Carpenter, Woolf, 1984, pp. 172 et seq.; Pieper, Vitt, 1996). The relative frequency of citations can be broken down by specific technological fields. Higher relative frequency indicates a relatively stronger technological position.
- Patent data can be analyzed to reveal how a firm compares with its competitors with respect to patent output and to patent quality (Ernst, 1996). Such technology benchmarking shows the relative technological position particularly if it is broken down by technological fields. Similar analyses could be performed regarding contributions to the scientific or engineering literature. Analyses of this kind are greatly facilitated by a citation index.
- Technologies that have not yet found an application can be traced with respect to certain performance parameters over time. It is often assumed that this will lead to typical S-shaped curves. Brockhoff (1994, pp. 137 et seq.) gives a critical review of this approach. A nice example is the graph showing the temperature at which the superconductivity effect can be observed over time with different materials (Ayres, 1988, pp. 87 et seq.). A company should define

its position on the curve relative to its competitors. Data can come from an analysis of literature, such as conference proceedings, journal articles, internet messages or private conversations. Earlier case studies have shown that a surprisingly large amount of the state of the art of individual firms is on display in various publicly available sources, although the company may endeavor to secure confidentiality.

Let us now turn to the identification potential (IP). Again, only a few examples that help to recognize this potential will be presented:

- The functions to be performed should be addressed in a mission statement for the research department. This is particularly relevant with respect to building up an identification potential. Identification of external technological knowledge should be an ongoing activity, even if it is organized by projects. If projects are started on an ad hoc basis - that is, after recognizing a particular need - it is likely that they will come too late to secure an optimum lead time for the initiation of further activities making use of the knowledge. This can be concluded from the results of empirical research on the organization and effectiveness of technological intelligence in German firms (Lange, 1994, pp. 239 et seq.). Institutionalized systems of technological competitor intelligence scan a broader spectrum of technologies and competitors, they support the usage of the information gathered, and they are considered to be efficiency-improving and effectiveness-enhancing. Lange (1994) also identifies a deficit in the application of sophisticated methods of analysis and a danger of rating 'formal' information sources more highly than 'informal' sources, without having much reason to do so. Many variables interact with technological intelligence functions in firms when it comes to explaining their effectiveness. It is of special interest that empirical research has found that even for those business units meeting the criteria of 'defenders' in the Miles and Snow typology, a positive influence of monitoring technological advances can be established (Dvir, Segev, Shenhar, 1993, pp. 155 et seq.). Furthermore, task-environment changes appear to be more salient than general-environment changes in the design of scanning systems. "Those that face greater task-environment change use a wider scope of scanning, scan more frequently, and assign their top management teams greater responsibility for scanning. Such designs are effective because of greater need of these

organizations to respond quickly and appropriately to environmental changes" (Yasai-Ardekani, Nystrom, 1996, p. 201).
- From time to time research managers should ask themselves and their project leaders questions such as: How often have we been surprised by external research results relating to our fields of interest? What is the time lag between the first publication of an important new item of technological knowledge and our organization taking notice of it? What can explain that time lag, and which company activity contributes the major share to it? Are we tied in to the network of primary contributors to research in the fields relevant to our company? Answers to these questions may be compared between organizations or with responses to earlier investigations of the same items.

A different perspective needs to be adopted to address the absorptive potential (AP) Here we may investigate:
- Is this potential included in the mission statement? Remarks made above may be repeated here.
- Research management should investigate whether results from technological intelligence services are not receiving attention or are remaining unused, although they may be of potential relevance. If such behavior is observed, it could indicate the presence of the 'not-invented-here'-syndrome (Katz, Allen, 1982, pp. 7 et seq.; Mehrwald, 1996). It could also mean that the external knowledge contains substantial amounts of tacit knowledge, or that the internal researchers lack particular capabilities that would enable them to use the external knowledge. Therefore, the reasons behind a possible weakness in the absorptive potential need to be studied in detail.
- Co-citation analyses may reveal whether a firm's researchers cooperate with external peers in publications. In general, co-authorship can only be achieved if each one of the authors contributes significantly to the joint work.

The development of the absorptive potential is of particular relevance if the company engages in cooperative research. The funds spent on this are used unwisely if they reflect only the ability to identify the most pressing issues for the firm, without integrating the knowledge that may originate from the cooperation. Job rotation or exchange programs with participating cooperation partners indicate the ability to transfer new knowledge into the firm. Representation on a program committee of the cooperation signals an ability to select most promising fields of study.

Let us now look at indicators of the inventive potential. As this is the primary function of the research group, we should not expect it to be missing from the mission statement, if such a statement is in force at all. We may thus examine additional criteria that could be discovered and evaluated by peer group audits:
- Is the research group able to demonstrate innovative results? Alternatively, its results might only replicate what has been found by others or they might be of less value (because more expensive material is used, more elaborate processes are needed, transfer to larger scale production is not feasible, etc.).
- Does the research group spend too much time achieving its results? If so, a detailed analysis of the reasons for this should be initiated as a basis for further action.

It is an interesting question whether peer group audits are preferable to audits performed by supervisors. Domsch and Jochum, who also suggest an organizational set-up for peer group audits, conclude on the basis of the literature available that
- "peer ratings are less affected by an overall impression of performance than supervisory ratings
- peer ratings are more suitable to discriminate between professional competence and other criteria of appraisal (willingness to perform or skill in communication) than supervisory ratings
- peer ratings do not show any more 'leniency' than supervisory ratings
- where raters participate in the selection of rating criteria and the construction of rating scales, peer ratings are better suited to discriminate between different aspects of performance than in cases, where a set is merely imposed on the rater
- peer ratings are not any more affected by friendship/enmity than supervisory ratings" (Domsch, Jochum, 1983, pp. 143 et seq.).

Although these results are derived exclusively for research laboratories, they show strong indications that peer audits have a methodologically sound basis when performed according to certain organizational rules. Supervisors who want to avoid personal biases may prefer applying 'objective' criteria in the evaluation, such as publication or citation counts. If certain precautions are not taken, citation analyses may have severe shortcomings even for the evaluation of academic basic research (Martin, Irvine, 1983, pp. 61 et seq.). Although more sophisticated approaches have been developed in the meantime, still a major shortcoming of such measures is the fact that corporate basic researchers may experience considerably less pressure to publish and be cited than their colleagues in academics.

The internal transfer potential (IT) may be evaluated with the aid of the following considerations:
- Is this potential included in the mission of the research laboratory? If not, researchers may develop the 'we are the champions'-syndrome or the 'not appreciated there'-syndrome. They may consider themselves to be the masterminds of the firm, expecting downstream departments to come and fetch results from them. This does not function very well in reality, if only because the other departments must be aware of the research results and they need to have access to the tacit knowledge in order to make the research applicable.
- Indicators of the potential may be collected by posing questions such as: How long does it take for research results to be integrated into development projects, and then finally into new products or processes? How long does it take for top management to notice any results from the identification potential of the research group? Are research results applied more quickly by other firms than by our own company? Do downstream departments reject research results, and if so, why does this happen?

Similar evaluations may be applied to measuring the potentials of downstream departments. Of particular importance is the answer to the question of whether these departments or groups are receptive to the results of a research department or group. They need to cultivate 'antennas' that can spot results of potential interest. This is important because a new research result may have broader or different applications than those imagined by the research staff. However, the specific criteria for the identification of potentials in downstream departments vary among those departments. We thus refrain from presenting these individual criteria as they can be found in the relevant literature on marketing, manufacturing, finance, etc.

It is important to recognize that diagnosis of the particular strengths or weaknesses of a research group in building potentials is possible. Organization controllers or the top management are held responsible for the initiation of such diagnostic analyses. Data collection for these analyses should be organized as an ongoing activity, as data collection in retrospect will include certain biases, for instance the rationalization of past behavior. The results of such analyses can be used for strengthening the research function. If this objective is communicated, it should help in winning the support of the research personnel for data collection.

The analysis - if performed over longer periods of time or in retrospect - can uncover whether a laboratory improves its contributions to competitiveness by becoming a 'star' or not. The manager of a social-science research institute of the automobile industry observes that in the beginning of his work, the institute was best described as a group of 'hermits that have to be visited' (see Table 9). The researchers then changed into 'successful research, with the problem of autism'. Fighting some remaining 'not-invented-here'-syndrome (case 16 in Table 11) they now work hard to establish themselves as stars. This illustrates that the cases are real. For any laboratory it is helpful to identify the present situation and the path taken to reach it. In the case reported here, pressure from shareholders and a switch to only partial institutional funding from the overhead of the shareholders initiated the moves taken and supported mixed financing of projects.

9.3 Suggestions for building potentials

Referring back to tables 8 through 12 we find that, in the majority of the cases, the contribution of research is questionable (as indicated by "?") or even absent (-). Particularly if such cases have been identified on the basis of the diagnostic analyses just mentioned, management will be interested in specific suggestions for strengthening research functions and thus building research potentials. These suggestions concern very diverse fields: improvements of interface management, strategic use of human resource management, better communication management, with particular emphasis on newly developed information and communication technologies, and investments in laboratory equipment or buildings. This can support creativity by giving opportunities for new types of tests or experiments or by relocations to enhance communication (see chapter 10). All of these fields are interdependent, but not necessarily mutually supportive. For instance, hiring highly creative researchers to grow the creative potential may reduce the internal transfer potential, and could therefore require further actions.

Consider another example. Supporting the use of modern communication links may enhance creativity only if a certain level of diversity of information inputs is maintained. This level is endangered if powerful search machines are employed that help scientists to screen information bases. "As quickly as information technology (then) collapses barriers based on geography, it forces us to build new ones based on interest or time. Ironically, global communication networks can leave intact or even promote partitions based on speciality, politics, or perceived rank, divisions that can matter for more than geography" (Van Alstyne, Brynjolfsson, 1996). Little seems to be known on how to combine breadth and depth of information intake to achieve most effective problem solutions. Operational optimization rules for choosing among the different suggestions can be developed at best in an iterative fashion based on careful analyses of past learning experiences.

It is impossible to name all suggestions that appear to be feasible. To illustrate our arguments we shall outline only a few of them:

If the relative technological position is seen to be low (TP = 0), the consequences may be one of the two following:

– Research capacity needs to be strengthened. This includes a review of the project and program portfolios, and possibly a change in the research personnel as well. It should be recognized that there may be a considerable time lag until these measures have an effect. This time span might be reduced if cooperations with strong partners could be initiated. This may require additional funds for the hiring of new personnel, as the outside partners could doubt the quality of the presently employed staff and therefore hesitate or reject to cooperate.

– If financial resources are extremely scarce or if the first alternative consumes too much time to become effective, research could be dissolved and the company could choose to shape an imitation strategy. This would have to be built up on strong environmental scanning potentials and a strong development or application engineering department (Schewe, 1992, pp. 226 et seq. with further success criteria).

University research results are better known to technology managers than those of private research institutions (Eggers, 1997, Tables 18, C8). While the former institutions generally publish freely, the latter tend to specialize more on contract research, which can lead to publication restrictions. This suggests that the effort needed to build up a high identification potential can vary depending on the sources of new knowledge. If these sources move to institutions that observe more strict secrecy, or if secrecy comes to be observed more strictly as a consequence of a policy change in a group of institutions, efforts to build up an identification potential may need to be increased.

Assuming that the identification potential ($ID = 0$) is at a low level, we would suggest the following actions:

– Install or revitalize a technological intelligence function close to the research group. Schering AG reported at a recent conference that one of its three 'Technology Offices' performed 147 non-confidential evaluations, 57 confidential evaluations, it supported 39 science-business sponsorships, it had 26 so-called second stage discussions which resulted so far in six negotiations and three completed deals. This remarkable level of activity was reached within nine months only.

– Plan an information session on this function for the company's top management to raise its sensitivity to new technological developments that could affect the company, as well as to win support for further long-term technological intelligence projects. Such projects could also serve as a first assignment for newly hired staff from universities. This would make use of their advanced knowledge, their relations with academic or other research institutions, and would serve to socialize them within the company. Reiss and Balderston even take the extreme position that "you do not have (researchers) because you expect to apply the results of their research. You have them for their ... ability to perform the advisory and service functions" (quoted by Berthold, 1968, p. 180)*.

– Discuss the relevance of traditionally used sources of technological knowledge. High fluctuation of scientists between institutions and changing roles or missions of publicly sponsored research institutions call for a review of traditional relationships. Perhaps more so

---

* (original publication: The Usefulness of Scientists, International Science and Technology, May 1966, pp. 38-44)

than in other 'purchasing' relationships, that of purchasing external technological knowledge appears to be more centered around creative personalities than around institutions. Publicly funded research and development institutions perform a wide spectrum of missions from fundamental research to development, application engineering and testing (for Germany see: Meyer-Krahmer, 1992, p. 427; for Japan see: Science and Technology Agency, Japanese Government, 1995, p. 130; for the U.S. see: National Science Board, 196, Chap. 4). It may be assumed that each institution tries to attract the type of personnel that corresponds best to its mission. In this respect, it is then interesting to observe which of the institutions produce the more relevant research for industrial research laboratories, and which is the more accessible. Both aspects are important because relevant work needs to be accessible to leave an immediate impact on the firm. Figure 15, taken from empirical research by Eggers (1997) in Germany, indicates that German universities continue to be regarded as producers of highly relevant research that is also quite accessible. The ranking order of research institutions on both scales is almost the same, with only small differences among institutions at the bottom of the scales. These rankings are further supported by an identical ranking of the frequency of cooperations between firms and the respective external institutions that perform research work (Bardenhewer, 1996a). According to the same study, U.S. respondents rate universities higher than private research institutes with respect to relevance and accessibility of their work. Interview results from Germany and from Japan indicate that a majority of managers expect university research to become even more relevant and accessible in the future, with some interesting exceptions (Bardenhewer, 1996a): The respective national universities are not necessarily regarded as conforming to these expectations, particularly in Japan a minority of European managers expects high budget pressure on European universities to reduce accessibility of their research, as they may try to increase secrecy in an attempt to improve the appropriability of income from research. If this would come true, the traditional role of the European research universities would be changed dramatically.

Fig. 15: Relevance and accessibility of external sources of research in Germany

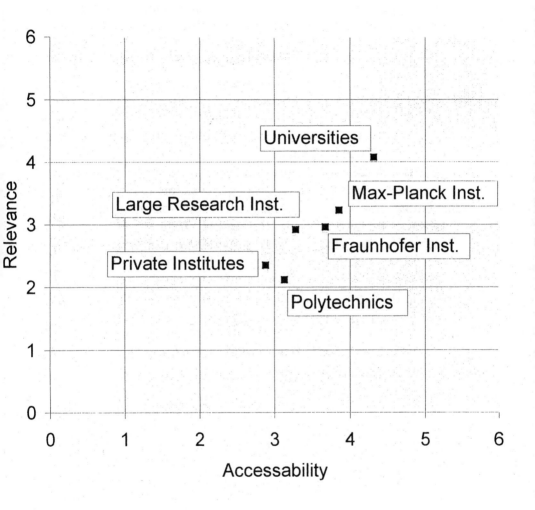

Source: Eggers, (1997)

Problems in interorganizational transfer of new knowledge can be attributed to all of the institutions (and their members) involved, as well as to the channels used for transfer. Eggers' (1997) study shows that understanding the results to be transferred is less of a problem than their applicability; in fact, the latter is the second most important problem area from the point of view of both German and U.S. technology managers. This is related to a low level of the absorptive potential (AP = 0). Such a low level may call for the following activities:

- Support the role of technological gatekeepers or boundary spanning individuals (Tushman, Scanlan, 1981, pp. 83-98; Katz, Tushman, 1983, pp. 437 et seq.). This will require early identification of such individuals and helping them to grow into their function. Until now, the acquisition of the gatekeeper role may be described as having been a social process developing almost unnoticed by formal human resource development activities. This may explain why it takes an average of 13 years (Meyers, 1983, pp. 199 et seq.). Contrary to the work by Katz and Tushman (1983), a German study discovered that researchers benefit more from the presence of gatekeepers than development personnel (Domsch, Gerpott, Gerpott, 1989, pp. 115 et seq.).
- Reduce the 'not-invented-here'-syndrome. This requires a more detailed look at this psychological construct, that can arise both at the individual level and at the group level (Mehrwald, 1996). As the syndrome seems to intensify with the age of a group, renewals of the group structure by inserting new group members could be a counter-measure. Certainly, this needs to be balanced with the beneficial effects of ongoing collaboration between group members. It is not easy to measure the net effects of changing group composition.
Schering Corp. in Berlin has initiated three 'offices of technology' in different countries that are given the mission to help integrate better external and cooperative research and to reduce potential 'not-invented-here'-syndromes. It remains to be studied which instruments prove to be most successful in this respect.
- Train researchers to integrate external technological knowledge and rotate personnel with outside producers of research results. In Japan, the initiation of twelve programs supporting the exchange of personnel between industrial and other research institutions in-

dicates recognition of a particular weakness in this area (Science and Technology Agency, Government of Japan, 1995).
– Establish an internal knowledge base by initiating projects that match the external knowledge. In interviews with European and Japanese technology managers it was found that accessibility of outside research appears to be significantly easier when a firm conducts related research of its own than when it does not (Bardenhewer, 1996a).
Still, Japanese technology managers are significantly less convinced than their European counterparts that internal research provides a lever for integrating external research results (Bardenhewer, 1996a). Whether this evaluation is due to different modes of transfer (for instance more frequent involvement in cooperations with other firms rather than universities in Japan than in the U.S. or Europe (Kolatek, 1990, p. 193), information exchange within groups of common origin cannot be determined.

An early study identified the following obstacles to successful transfer: different attitudes between institutions, the 'not-invented-here'-syndrome in the receiving laboratory, problems of personnel transfer, physical separation of groups, and a lack of control at the downstream unit. These items have been presented in ranked order of their perceived importance (with the exception of the 'not-invented-here'-syndrome, which was added to the list only later) (EIRMA, 1988, p. 27). More recent empirical research has tried to uncover mutually independent influencing factors by applying factor analysis to a similar list of reasons. This indicates that problems of internal transfer result from communication difficulties (difficulties in understanding results, the willingness to accept results, the awareness of the results, the geographical distance between the groups) and from coordination or effectiveness difficulties (differences in planning horizons, insufficient applicability or market orientation of basic research results) (Eggers, 1997, Tables 16, 17, C7). The study's German and U.S. respondents cite the lack of coordination with respect to planning horizons as being the most severe problem. The communication difficulties are perceived to be somewhat more important by the U.S. managers than by the German managers. Thus, there is good reason to improve the internal transfer potential in many firms.

Better alignment of planning horizons may be achieved by top management or a powerful controller. However, more measures may be necessary:
- As before, the external transfer potential should be addressed in a mission statement. Before it can go into the statement, the importance of this point to the firm and to the laboratory needs to be communicated to researchers.
- Researchers need motivation to support transfer activities. For many of them it is more rewarding to generate new knowledge and to discuss this with their peers than to evaluate potential development problems that might result from it.
- The cure of the trouble may be less expensive if receptiveness for research results in downstream departments can be heightened. This may be achieved by representing these departments on an advisory council, by inviting possible 'customers' to presentations of results, or by reducing 'not-invented-here'-syndromes at the receiving units.

The necessity to improve the internal transfer of research results has been a topic of recurring interest in the management literature (Rubenstein, 1957; Goldman, McKenzie, 1966; Rubenstein, Barth, Douds, 1971). Communication problems are often considered as the major obstacles to more effective transfers. Consequently, most suggestions for improvement explore possibilities for enhanced communication. These suggestions can involve multi-item constructs, such as 'intergroup climate', type of 'task interdependence', and 'work-related values' (Rubenstein, Barth, Douds, 1971). While the second construct cannot be changed easily, changing the other constructs can involve issues of human resource management, team building, or development of a joint company vision. It is seen easily that each of these suggestions may be supported by many specific actions.

We would like to mention only a few examples of such actions. Apart from effectiveness problems, a low level of the internal transfer potential (IP = 0) may require actions similar to the ones mentioned when discussing the development of the external transfer potential. They may, however, address different persons. The following are possible plans of action:
- Developing or strengthening of the role of internal stars. These are persons who play a role similar to gatekeepers or boundary spanning individuals, albeit in a strictly internal environment (Tushman,

Scanlan, 1981, pp. 83 et seq.). They could help reduce communication difficulties.
- Reduction of internal communication syndromes, such as the ones that were identified above. Furthermore, the predominance of written information exchange needs to be supplemented with other forms of communication that are 'richer' and more 'vivid'. It is interesting to note yet another result from the Japanese-European comparison of research management: Japanese rely more than Europeans on presentation media (video conferences, conferences, videos) and on the internal transfer of personnel in either direction; they use less information workshops, sample distribution or e-mail messages (Bardenhewer, 1996a). These results should be interpreted individually. They may interact with other variables, such as the degree of internal customer representation on advisory boards to the research laboratory or individual projects.
- In a study on central research laboratories team research was mentioned as the most common method of bringing research methods to market (Bosomworth, Sage, 1995, p. 39). Although this may not apply to all types of projects it could be helpful for some of them.
- Job rotation between research and downstream departments in either direction, as mentioned in the preceding paragraph.
- Inclusion of representatives from downstream departments in advisory or review boards to the research group. This could help to align planning horizons on all sides.
- Support project funding in the way discussed in Chapter 5, above. This would necessitate a discussion of time schedules, and also addresses the effectiveness issues.
- Establishing a formal transfer process, because it reduces transfer time compared with informal processes. Two reasons are given for this suggestion:
"1. A formal transfer procedure may be the result of lessons learned from previous transfers. If so, it eliminates the learning curve each time a new transfer is attempted. This is particularly valuable as a person in charge of one transfer may never be responsible for another transfer.

2. The existence of a written process, no matter how meager, provides an early agreement between the business unit and central research. It can eliminate false starts and define responsibilities that would otherwise take time to evolve" (Bosomworh, Sage, 1996, p. 39).
- Evaluation of distance between research groups and their major scientific contacts as well as their customer base.

Alcoa's researchers have repeatedly voiced a concern that they should not be located too close to operational units, as this may tempt these units to determine more application oriented research programs; research was seeking for a 'neutral' location that could both maintain contacts with the scientific environment and avoid getting too close to the 'fumes' of the works (Graham, Pruitt, 1990, pp. 128, 153, 155, 203). Here, the word 'neutral' signals that research was aware of the relevance of distance, although it lent more towards a shorter distance to science than to the 'downstream' operating units.

Deficits with respect to the inventive potential (IP = 0) may require radical personal or organizational changes, such as:
- Reorganizing research laboratories, for instance by integrating those that have become advanced development groups into the development departments of business units. This frees up research funds that may be used to initiate new groups.
- Transferring individuals to other functional departments once they have slowed down their creative thinking.
- Reviewing research equipment. It might turn out that with the equipment at hand it is impossible to keep abreast of new developments.
- Involving top management and the management of business units in strategic mission development for the research groups.

As one can see, some of the above suggestions (such as job rotation or representation on advisory boards) already involve the downstream departments. These departments' interest in research may not come naturally, and would therefore need to be developed or boosted. Gupta and Wilemon (1990, pp. 277 et seq.) show that the actions to be taken in order to bridge interfaces among marketing and R&D departments differ among marketing, R&D and top management. What can be advisable for one group, may not be feasible or reasonable for another. As before, we do not present specific suggestions here, as these would then have to include the focus of the different downstream activities.

In conclusion, we can see that the concept of research functions which lead to research potentials is indeed fruitful, not only in terms of individual research projects' funding, but also in defining necessary performance conditions for a research laboratory. The following figure (Figure 16) demonstrates the process that might be followed in analyzing the situation.

Fig. 16: A process view for improving the necessary conditions for research performance

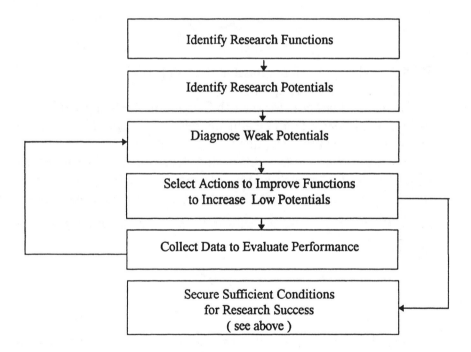

9.4 How much attention for each function?

We argue that potentials grow from continuous support of the respective research functions. This is a simple assumption, which does not consider any cross-effects between functions relative to a certain potential. More specifically, we speculate that more recent investments in a function have a more intense effect on building a potential. With an increasing distance in time the effect could be assumed to decline in a geometric fashion. Taking these assumptions together, we arrive at an expression that explains the current level of a potential by:

$$P_{i,t} = a_i \; x_{i,t}^{b_i} \; x_{i,t-1}^{b_i c_i} \; x_{i,t-2}^{b_i c_i^2} \ldots$$

with
$P_{i,t}$ the level of the i-th potential in the t-th period, i = 1, 2, ..., I and t = 1, 2, ...;
$x_{i,t}$ the level of support for the i-th function in the t-th period;
$0 < b_i < 1$ a parameter that explains the relative change of $P_{i,t}$ caused by a relative change of x_i; this is an elasticity;
$0 < c_i < 1$ a parameter that explains the decay of the original effect expressed by b_i for each additional step by which the distance in time between the functional support and the potential is increased.

Dividing $P_{i,t}$ by $P_{i,t}^{c_i}$, we can substitute our expression by the more simple formula:

$$P_{i,t} = \frac{a_i}{\frac{c_i}{a_i}} \; x_{i,t}^{b_i} \; P_{i,t-1}^{c_i}.$$

We would also like to secure that expenditures on different functions do not exceed the total research budget X_t in the t-th period. This leads to

$$\sum_{i=1}^{I} x_{i,t} = X_t.$$

Without loss of generality, we may assume $X_t = 1$. The objective is to distribute funds to the different functions such that the overall resulting potential or strength of research is maximized, i.e.

$$\max \prod_{i=1}^{I} P_{i,t}$$

and the budget constraint is not violated. This problem finds a simple solution, namely

$$x_{i,t}^{*} = \frac{b_i}{\sum_{i=1}^{I} b_i} \cdot X_t .$$

This means that the optimum support for the i-th function is determined by its elasticity relative to the sum of the other elasticities.

If we now assume that we are not only interested in the optimization for only one period, but rather for

$$\sum_{t=1}^{} \left(\max_{x_{i,t}} \prod_{i=1}^{I} P_{i,t} \right)$$

we find under the same constraint

$$x_{i,t}^{**} = \frac{b_i (P_t + c_i P_{t+1})}{\sum_{i=1}^{I} b_i (P_t + c_i P_{t+1})} X_t .$$

To simplify we assume that $P^*_{t+1} = \alpha P^*_t$, $\alpha > 0$. Then we have

$$x_{i,t}^{**} = \frac{b_i (1 + c_i \alpha)}{\sum_{i=1}^{I} b_i (1 + c_i \alpha)} X_t .$$

Analyzing this result we find that one should spend more on the i-th function if b_i or c_i get larger, and less if b_j and c_j, where j indicates the rivaling functions, get larger, all other things being kept unchanged. This means that those functions that contribute most to building the related potentials should also receive heavy contributions.

It would only moderate but not change the result in principle, if the individual potentials do not contribute equally to the aggregate potential P_t, but would have a weight attached to them.

The model is highly speculative, as until now it has no empirical support. This shows that much more research is needed to arrive at substantiated conclusions that could eventually help research managers to determine at least approximate levels of support for individual functions. Before finding uncovering such empirical information, we refrain from developing more complicated and speculative models.

10. On locating research

10.1 Qualitative analysis

After the identification of various research functions has been achieved, two related questions need to be answered: Should all functions be performed at the same location or not, and where should this happen? When industrial research was first started, the answers were relatively simple. All research activities were concentrated at one location, the choice of which depended very heavily on the mission of the research department. When the primary mission was to support technological decisions of top management, research was made a headquarters unit. When the function also included supporting development, which was typically located close to the main manufacturing operations, research was in turn located close to development. As the headquarters was very often located where major manufacturing occurred, research was here again close to the headquarters. If the mission stressed heavily the absorption of general scientific developments or 'freewheeling' creativity, research was located close to major universities or in other possibly creativity-enhancing areas. This could result in a belt of similar institutions around a core institution. The times when location decisions appeared that simple are long gone. Apart from a national location decision, in which the aspects of distance discussed above come into play, international research-location decisions are gaining in importance at present.

As mentioned above, Hitachi Ltd. has located research laboratories in many countries outside of Japan. It has been reported that of the 28 laboratories that Japanese firms set up outside of Japan between 1987 and 1992, 15 are charged with the mission to contribute to research. The same is true for seven out of nine laboratories in the pharmaceuticals industry. The laboratories are tied very closely to the scientific establishment in their respective field in the host country (Kümmerle, 1993, p. 24). In British laboratories of foreign multinational firms charged with basic research, one third of them collaborate with their scientific environment. In applied research laboratories this share is twice as high, while it is much lower for other laboratory functions such as product or process development (Pearce, Papanastassiou, 1996, p. 331). Contrary to Kümmerle's findings, the laboratories of Japanese firms do not play leading roles in this respect. This may perhaps be explained by the differences in perceptions on the degree of interaction between headquarters and local units.

Hoechst AG plans to spend 60% of its total research and development budget outside Germany, including basic research. In 1995, the company spent 53% of the total research and development budget abroad, allocating these funds to 90 laboratories (du., 1996, p. 21). This number is up from 12 foreign based laboratories in 1993[*]. According to a third source the company operates one central laboratory at its headquarters, three in the U.S. and one in Japan that are all charged with some or all of the research functions identified above (Neukirchen, 1996, p. 59). The central research laboratory performs service functions to all subsidiaries. In 1992 Siemens AG reported on laboratories in 23 countries outside Germany, employing more than a quarter of its total research and development personnel; the company mentions explicitly that some of these units cooperate on research problems with some of the best universities, including, for instance, Princeton University and the University of California at Berkeley[†]. Searching the business press, it is no problem to point to further examples, possibly involving even more foreign research activities.

Findings of a study of 84 British, German and U.S. companies indicate that 7.1% of their foreign research and development expenditures are allocated to basic research, and that 27% of the foreign laboratories perform at least some basic research (this becomes 27% of the funds and 92% of the laboratories, if applied research is also included) (von Boehmer, 1995, p. 67). In a cluster analysis of research missions, one group of laboratories that is called the 'cooperating basic research institutions' can be clearly identified. These make up 5.4% of all laboratories studied (von Boehmer, 1995, p. 87).

---

[*] Hoechst Japan, Ausbau der Forschung, Handelsblatt, December 13, 1993.
[†] Siemens Forschung, 47,000 Mitarbeiter in Laboratorien tätig, Handelsblatt, January 17, 1992.

These laboratories are mostly globally oriented and not locally. They differ from at least some other types of laboratories with respect to:
- a greater degree of financial and legal freedom
- more market specifics, which includes such items as monitoring competition, presence of major competitors in the foreign market, proximity to lead users, or specific customer requirements
- attractiveness of scientific institutions, which is related to the proximity of such institutions, the possibilities for cooperations and/or the general technological dynamics in the host country
- specific management capabilities, such as the quality of the management in the foreign country, its initiative and its technological sophistication, the capability of the company to manage its global network and/or transfer new technologies (von Boehmer, 1995, pp. 92 et seq.).

From this we can see clearly that establishing research units in foreign countries requires very specific management capabilities, and that it appears to be dictated by environmental requirements. This would support the case-based observation of de Meyer and Mizushima (1989, p. 139), that setting up foreign laboratories "is typically accepted more with resignation than with pleasure". It is consequently of particular interest that headquarters' R&D managers perceive both the effectiveness and efficiency of the basic research laboratories in foreign countries more favorably than any other type of development laboratories (von Boehmer, 1995, p. 110). Even though this result may sound reassuring, it does not differentiate by the different research functions.

With regard to the issue of centralization versus decentralization, it is valuable to distinguish between the performance of one or more tasks and the decisions to organize that performance. This should be distinguished from decisions that initiate individual research projects, which according to our scheme in chapter 5 are not likely to be centralized. In that respect, four models may be considered:
- centralized decision making and centralized task performance, which would lead to one central research laboratory
- centralized decision making and decentralized task performance, which leads to the existence of different centers of excellence where the tasks are performed

- decentralized decision making and centralized task performance, which is a model situation for a research center that operates as a cost center or a profit center on the basis of individual projects supported by different stakeholders within or even outside the company
- decentralized decision making and decentralized task performance, which would reflect strong local units. These laboratories may be called local research units (Medcof, 1996)*.

As there is no perfect market for the organization of research in the company, total decentralization of organizational decisions does not appear to be appropriate. While this could minimize managerial problems at the headquarters, it also raises doubts regarding technological performance criteria of the more autonomous laboratories (Brockhoff, Schmaul, 1996). As it is obvious that the different research functions may not all be handled at the same level of efficiency in one location, the natural choice is to have many locations, but one decision making unit that is involved in the allocation of tasks among the units. This formulation allows decentralized research laboratories to develop initiatives leading to the accrual of more and other tasks than those performed hitherto. With respect to different fields of research, this possibility was called the 'split and grow' approach by one company, meaning that a group that splits off from the mainstream research of a laboratory and adopts tasks in a field in which no other competence center within the company specializes, is offered the chance of growing into a full-blown research center. As observed for Japanese laboratories in Europe, intense cooperative interaction of these laboratories with the local scientific environment tends to increase their autonomy. Headquarters then become interested in exercising greater control, unless the laboratories develop strong internal communication links with their potential internal 'customers'. It appears "that managers use internal linkages as a condition based on which local autonomy can be granted," which includes the autonomy to start up research initiatives (Asakawa, 1996, p. 31).

A correlation between research functions and location preferences is clearly demonstrated in a study of German industrial research labo-

---

* See also the "corporate technology units" as described in Ronstadt, 1978, pp. 7-24; a more differentiated analysis of organizational options is given by: Chiesa, 1996, pp. 7-23.

ratories (Eggers, 1997, p. 107 et seq.). Three clusters of laboratories have been asked to evaluate the importance of short distances to any one of three internal groups and three external institutions. Significant differences of importance are observed for only three of these (not for the distance to headquarters, applied research or development, markets). Short distance to universities and other research institutions is of significantly greater importance to those laboratories which seek the relatively highest levels of innovation and which put strong emphasis on the identification potential, the absorptive potential, the image enhancement and the attractiveness for hiring new researchers. Short distance to internal manufacturing is of significantly greater importance to laboratories which seek better understanding of presently applied techniques and improvement innovations. No significant results have been achieved for a third group of laboratories with less well-definded potentials. This illustrates and supports our view that research functions and potentials are key to understanding research management.

Summarizing the casual observations we find that the allocation of tasks (which are functions with respect to particular fields of technology) to possible laboratory sites should be guided by the following criteria:
– the cost of performing the given task, alternatively called the organizational cost
– the cost of transferring necessary inputs into the unit
– the cost of transferring the unit's output to its customers.

The second group of cost may be incorporated into the first. We are then left to consider the production cost of new knowledge (first and second groups), and the transaction cost of new knowledge (which refers to the third group). The sum of the two should be minimal for a given task. Unfortunately, this cannot be easily determined for at least two reasons. First, many tasks are interdependent. The innovation potential may be greatly enhanced by a well planned identification potential, such that the cost of two separate projects in these fields may need to be considered simultaneously. Because laboratory sites may not be charged with performing only one type of these projects, difficult problems in evaluating the cost as a basis for decision making arise. Second, alternatives exist for influencing the cost. Earlier, it was argued that research on organizational interfaces has shown that transaction cost depend on the distance between the units

that need to exchange information, goods or resources, where distance is not exclusively thought of as reflecting a physical or geographical meaning. "If locational decisions increase distance, [transaction] costs tend to rise. This calls for either one of three decisions:
- substitute organizational cost by transaction costs if these promise to be lower... ; this could involve a change of strategy ...
- move the 'neighboring' function (or some part of it) towards that function that caused the organizational cost increase in an attempt to decrease [the total] cost again; this would involve a change of structure;
- try to influence competitive conditions such as to make it easier to bear the cost. This, however, seems to be an option that we should not follow up" (Pearson, Brockhoff, von Boehmer, 1993, p. 256)*.

Data from the U.S. lend implicit support to the relationship between distance and cost. The proportion of funds given by industrial firms to various universities for basic research varies with two major variables: one is the distance between the research laboratories in industry and in universities, the other is the quality of the university as measured by the National Academy of Sciences (Mansfield, Lee, 1996, p. 1053). The relationship is almost negatively exponential, with some interrelationship with firm size. It is believed that "the more fundamental the research is, the less distance will matter because fewer and less intensive interactions between firm and university personnel will be required, and because the technical expertise of the faculty will be of greater and greater importance" (Mansfield, Lee, 1996, p. 1055). While these arguments reflect currently observed behaviors, it is not obvious that this is optimal behavior. Thus, more collaboration between particular universities and industrial firms might be fruitful if more industrial laboratories were moved closer to the respective universities. This could be of special interest when it comes to absorb the external knowledge and to use it creatively.

A fourth option could be
- to reduce transaction cost by employing new communication technologies (Pearson, Brockhoff, von Boehmer, 1993, p. 257).

---

* In this text what is called organizational cost was labeled above production cost.

Considering the fact that a major share of the production cost of research is personnel cost, and that this cost is usually much higher than the transaction cost, this could be the starting point for planning research locations. The questions to be answered would be where to find the kind of personnel needed to perform the identification of new knowledge or the creation of new solutions, and at what price they are to be found. Next, it may be studied whether this personnel can be transferred from their home base to some other location, possibly the headquarters. This may be easy in some cases, as moving to the headquarters location adds benefits to the job of the researcher. In some other cases this may be extremely difficult, if the researcher perceives the possible move as detrimental to utility that is derived from the job. As such considerations can be calculated in money terms, they can serve as a basis for the location decision. The next step would then be to add the transaction cost. These are more difficult to evaluate in money terms, particularly as many qualitative issues influence the level of these costs. Taking everything together, Chiesa (1996) appears to have found that there may be reason to practice greater dispersion for research laboratories than for development laboratories, partly because the identification potential can best be built close to where the original scientific knowledge is produced. In terms of cost, this implies that the cost of transferring the necessary input into research is kept low if a unit is close to where the knowledge input originates, and also that the cost of transferring the unit's output to its customers is relatively low. If the unit were moved closer to its internal customers, the first cost item would increase substantially, while the last item would not be decreased enough to compensate for the first effect. Similar reasoning, possibly substantiated by numbers reflecting the cost, should be made for all research functions, and by individual research areas.

10.2 Modeling the location decision

In a very simple model it is possible to illustrate and sharpen some of the features that determine the location decision. Let us assume that research is one of only two factors of production, and without loss of generality we may even assume that it is combined multiplicatively with a constant amount of other factors. Research is composed of two activities or types that are mandatory, using external research results and creating original results. A maximum number of m external results could be used, if the distance between the research group and the only place which produces the external results could be reduced to zero. However, if the distance grows to $d_x > 0$, the number of usable results decreases to md_x^{-a}, $a > 0$, due to a decreasing frequency of mutual communication exchange. The return which originates from these interactions is derived from applying an elasticity $s>0$ to the number of interactions, resulting in a partial contribution to return $(md_x^{-a})^s$. The cost per unit of communication is K. Set-up cost is neglected at this stage, as the number of laboratory sites is not under consideration.

Similarly, creating original results benefits from interaction with downstream departments, and a maximum of n such results is feasible. As the distance d_y between the laboratory and the internal, downstream departments grows, we have nd_y^{-b}, $b > 0$, interactions which result in a partial contribution $(nd_y^{-b})^v$ with elasticity $v > 0$ and which cause cost of Knd_y^{-b}.

The shortest total distance D between the scientific environment and the downstream departments is $D = d_x + d_y > 0$.

Putting the details together into a profit function (π), we find

$$\pi = (md_x^{-a})^s (nd_y^{-b})^v - K(md_x^{-a} + nd_y^{-b}).$$

The company wishes to determine d_x and d_y such that profit is maximized. After taking first derivatives and simplifying we find an interesting result:

$$\frac{s}{v} = \frac{md_x^{-a}}{nd_y^{-b}}.$$

If s grows relative to v, it is obvious to expect that relatively more interactions with the scientific environment are optimal. As m and n were assumed to be given, this could be achieved by decreasing d_x and by increasing d_y. If s and v were given and, for instance, relatively more contacts were deemed necessary, it is clear that the optimum would lead to moving the laboratory closer to its internal customers. In general terms, the result says that the distances should be chosen such that the proportion of the communication frequencies equals the proportion of the elasticities. This result can be extended to more than two types of communication links.

Simplifying further by assuming a = b we can see that equal distances d_x and d_y require that s : v = m : n or that the relationship between the elasticities of science contacts versus internal contacts equals the relative frequency of the maximum number of contacts.

Assuming that the two types of communication are substitutive, we should add the first two components of the profit function (above) rather than to multiply them. This has an interesting consequence for the optimization decision. We find

$$\frac{s}{v} = \frac{md_x^{-a}}{nd_y^{-b}} \div \frac{(md_x^{-a})^s}{(nd_y^{-b})^v}.$$

Here, the distances have to be chosen such that the proportions of the frequencies of communication per unit of the results that originate from them equals the proportion of the elasticities. Again, more than two types of communication or mixed modes of result generation could be considered.

Furthermore, cost could not only depend on the frequency of communication, but on a frequency weighted by distance, to reflect travel cost. This leads to the following optimality condition:

$$(md_x^{-a})^s (nd_y^{-b})^v \left[\frac{v}{nd_y^{1-b}} - \frac{s}{md_x^{1-a}} \right] = K \left[\frac{b-1}{b} - \frac{a-1}{a} \right].$$

This could be evaluated numerically. However, if we assume that the 'decay rates' for communication in relation to distance are equal (a = b), then we get

$$\frac{s}{v} = \frac{md_x^{-a}}{nd_y^{-b}} \cdot \frac{d_x}{d_y}.$$

Compared with our first optimality condition, here distance-weighted communication frequencies need to be equated with a quotient of elasticities.

What can be learned from these little exercises? The following points may be noticed:

- The unit cost of communication is irrelevant to the decision if either the rate by which communication frequency decays as distance increases is the same for all types of communication relations or the cost is different for different types of communication.
- Then, distances are determined by their elasticities with respect to the result produced from the new knowledge and a weighted quotient of the maximum frequencies of communication.
- The weights are determined by the cost structure or by the conditions that describe the generation of results from the communications. The weight is one, if the cost is only dependent on communication frequency and the interaction of different types of communication is strictly multiplicative.

Thus, modeling the location decision can add structure to the qualitative reasoning on location and add to its precision. It also shows what kind of information is needed to make a rational location decision.

Having decided on the location of one laboratory it is possible to calculate the resulting profit and compare it with current raw profits $m^s n^v$ that would arise if two laboratories were located at zero distance from their respective partners of communication. However, this solution would entail a cost of communication between the two sites and additional investment in infrastructure that could be used by both laboratories if they existed in one location and which now needs to be doubled. Here, the evaluation could be done numerically on the basis of a capital budgeting model. It should be kept in mind that we have argued in Chapter 8 that over-decentralization should be avoided.

11. Conclusions

Business has become increasingly technology-dependent. The generation of new technological knowledge as a source of competitive advantage has become a bottleneck in many firms. This could reflect a long term trend (Brockhoff, 1996a). A natural response to this trend might be to strengthen research and development activities both inside and outside the firms. With respect to its long term potentials, research could be expected to take a prominent position in a program for growth and better future competitiveness. But on the contrary, as often before, recent reactions to environmental and competitive pressures as well as the rediscovery of shareholders' claims, particularly in the U.S., have cast industrial research and development into disfavor. Research, in particular, has come under severe pressure, now having been substantially reduced in many firms and entirely eliminated in others. While research directors have been fighting against this development, they have found few weapons with enough power to convince seasoned managers of other functional areas, controllers, outside directors or CEO's. Until now, management science has had little to offer that would help support the views of either side in these battles for scarce resources. The specific characteristics of research cause this activity to be pushed out of the annual planning rounds: these favor more regular business activities having promising results in shorter time, a higher probability of success, and much better chances almost fully appropriating the results that follow from the input of various factors of production. Although case histories have been presented in recent years that show the detrimental effects of eliminating research or of fluctuating research activities in business firms, this material is either unknown to many managers or it is unable to convince them of the possible benefits of industrial research programs.

We have tried to collect evidence on the potential long term benefits of industrial research. Furthermore, we have systematically explored different functions of industrial research that go beyond the

creative development of new knowledge that may serve as a basis for future new products or processes. The functions identified in earlier research as well as in a newly performed empirical study contribute to building research potentials. We have found five primary and three secondary potentials (see Figure 8). Knowledge of the potentials is of substantial importance for managers:
- The potentials can be used as a stimulus for thinking about possible sources of funds for research projects and most likely sponsors. Doing this systematically may broaden the basis for research financing. A 'funding form' (Figure 10) was developed that is based upon this idea. It was tested successfully in a number of large European electronics companies.
- The potentials cannot substitute for one another completely. Instead, a balanced presence of all potentials appears to establish a necessary condition for research success. We think that it is extremely helpful to be aware of this. We have, therefore, developed (in chapter 9) a taxonomy of necessary success conditions for research.
- In order to make this more applicable, we have also discussed indicators of missing research potentials that could spell trouble. If these indicators diagnose that particular potentials are lacking, these should be attended to and developed in a systematic manner. We have provided suggestions for the building up of potentials. While these draw on a substantial body of literature and earlier experiences in research environments, they are not considered to be exhaustive. The presentation should be used as a stimulus for the development of further suggestions.
- A flowchart summarizing the steps to be taken to organize the process of establishing necessary conditions for research success was developed and is presented above.
- The identification of different research potentials sheds light on yet another problem. It is very plausible that not every potential needed to make research a success for a company can be provided efficiently and effectively at only a single location possibly the headquarters of the company in question. If a multitude of research sites is chosen as a response to the differentiation by functions as well as a specialization by areas, the external and internal transfer potentials become of particular relevance in securing research success. Location decisions must include consideration of the transaction

costs related to these transfers, as well as the cost of producing new knowledge in different places. Task interdependency can make these decisions extremely difficult.

The research potentials, as applied to project funding and management decisions, provide a bottom-up view of the research planning process. This view needs to be balanced in a top-down perspective. As discussed above, this entails at least two particular aspects that can be used in research planning.

First, research success is a consequence of meeting necessary as well as sufficient conditions. The sufficient conditions, which should not be seen in a deterministic way, revolve around the proper choice of research topics. Certainly, nobody can foresee the potentials of new scientific knowledge, and therefore the sufficient conditions cannot be derived as a set of rules for immediate project selection, possibly using frameworks from financial analysis. We think that three steps need to be discussed. It is necessary to determine whether a substantial science base for business relevant technologies can be discovered. If so, projects should be defined such as to support or to develop core competencies (in the sense that this term is used in the business literature). Ongoing research should be evaluated to eliminate overperfectionism and unwanted bootlegging, as these activities employ resources that may be put to better uses. If research directors miss out on this point, they may lose credibility both with the general management and with their research personnel. Saving funds for better uses may be the less important result of a research director's own controlling activities as compared with the motivational issues just mentioned.

Second, top-down budgeting referring to research contributions generating new business may use the concept of elasticities, which is of very substantial importance in other business areas. It is true that estimating sales elasticities of research, perhaps the ultimate goal, is extremely difficult. However, econometric research can establish such elasticities from cross-sectional data in an industry or from time-series data in a firm. Such results may be used as a guide by research directors. They could then try to work on correcting the estimates by using Delphi-styled interview schemes to elicit perceptional elasticity estimates. This procedure has been described and tested. Certainly, further applications need to be developed and communicated to build a sense of the reliability of such estimates. It cannot be expected that the estimates will be fully reliable and valid from the outset. If experience does lead to better estimates, research directors gain a very powerful instrument to be used in discussions on budgeting vis-à-vis other business functions. It is for this reason that we advocate its further development.

On this basis, some additional insights can be gained if the role of research is better understood. In modeling the contributions of research to future sales by using elasticities, it becomes clear that research depends upon external variables, most importantly the interest rate. This dependency is intensified by the time lag between research and possible sales. A long time to the market and a high interest rate will reduce the share of research spending. When interest rates go up, pressure to perform more short term projects increases if the share of research is to be left unchanged. An understanding of such relationships is very valuable to research directors if they have to argue with finance representatives. Although neither side may employ the model as such, the market rate of interest can induce behavior conforming to the model results.

In summarizing the concepts developed here or elsewhere for purposes of research and development budgeting it is desirable to attempt a generalized integration. This we present in Figure 17.

The choice of a budgeting system in firms may be constrained by its operating environment. Major sponsors of research, partners in cooperative projects or even legislation can have substantial impact on the choice of budgeting systems. Another sort of environmental influences concerns the signals that codetermine data entering the budgeting process, as for instance the interest rate. Within a budgeting unit, whether this is a firm, a business unit or a similar area, the constraints and the signals may become relevant at an overall, high level or at a project-related, low level.

Following from this we can identify *top-down budgeting* procedures, where setting the total budget level based *on the level of elasticity estimates* would be a first major step. Then, the budget needs to be allocated to individual activities. If these activities are groups of projects with similar characteristics in economic terms, like research projects, elasticities for this sort of activities help us in determining budget shares that should be allocated to them. If activities are individual projects, this is not feasible. Due to the level of uncertainty involved it makes little sense to use a project related elasticity for planning purposes. Also, the individual project is usually considered as an entity that can either be undertaken or not. It is therefore difficult to conceive an x% change of the project budget in its relation to some output measure, while this is a valid notion for an aggregate of many projects.

Fig. 17: Views of the budgeting process

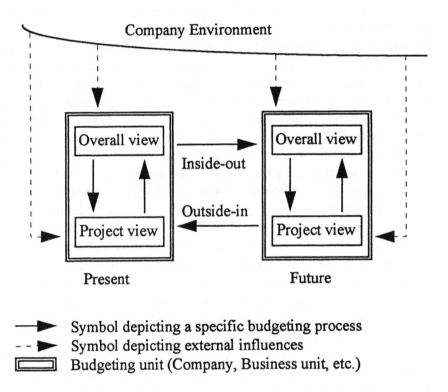

    Let us now turn to the project level. The *bottom-up budgeting* can build heavily on the different *research functions and potentials*. It employs these functions for broadening the funding base for projects by a systematic exploration of possible sponsors for projects and their functions. As always, top-down and bottom-up approaches need to be aligned.

    We expect that elasticity based top-down budgeting arrives at smaller budget levels than functionally based bottom-up budgeting because the elasticity estimates may not explicitly incorporate all research functions. The discrepancies are not of relevance if budgets are strictly project based. In other cases they should initiate discussions on the relevance of the research functions for the firm.

Assuming that all of this happens in the current accounting period an even more important question is how to plan budgets for a future period. As before, there exist two different approaches to this problem. First, present and past experiences can be used to project or to extrapolate into the future. We call this an *inside-out approach* which can again be based on *elasticities*. The draw-back of this approach is well-known. Major technological or market changes are not reflected in the past data and adjustments to goals of a budgeting unit may be extremely difficult. Therefore, this approach can be supplemented by an *outside-in approach*, which rests on future scenarios and goal levels. From these, one has to work one's way back to determine necessary budget levels. Here, concepts of elasticities and functional differentiation can be combined. Inside-out or outside-in approaches are tied to specific budgeting rules at the overall level. Strategic gap analysis lends itself easily to support an outside-in approach, while traditional capacity related budgeting procedures are definitely supporting inside-out approaches. Both of these approaches benefit greatly from simulating consequences of budgeting decisions on long-term company development or vice versa. This has been shown to work well for the development budgets, both in theory (Brockhoff, 1989) and in practical applications (Bender, 1997; Perlitz et al., 1996). It remains to be demonstrated whether and how specifics of research can be treated in exactly the same way or whether these require completely different tools. The experiences collected here seem to indicate that if functional requirements or objectives can be projected into the future it should be possible to use these for an outside-in simulation of research budgets as well. Certainly, specific relationships would have to be developed. This goes beyond the work presented here, and it is an important task for the future. Even discussion of advanced and presently available simulation software for research and development and its applicability to research environments faces considerable opposition in practice[*].

---

[*] Private communication.

Another consequence of modeling is the suggestion to explore whether research is mandatory for generating new business or whether it is supplementary to development. In the latter case relatively less would be spent on research and relatively more on development. This is a direct consequence of the relative size of the contributions of either activity.

Cost of internal transfer from research to development have been included in the model as a 'prototype' of similar cost items. It can be shown that the higher these transfer costs are, the less is spent on research. Therefore, it may pay to initiate organizational arrangements or programs to reduce transfer cost. This might be in the interest not only of the company as a whole, but also of the research director. At first glance, he might reject spending parts of his budget on easing transfer, adopting a hermit's view of his laboratories' findings: Let the rest of the company come and seek our wonderful results! Our model offers strong arguments for changing this behavior, as lower transfer cost can help increase the research share of total research and development expenditures. Similar ideas may be developed with respect to consequences of the 'not-invented-here'-syndrome.

Thus, interesting new results have been obtained that can help to better understand industrial research. Certainly, much more remains to be done. The present model could be developed into a fully dynamic model that sets out to explain growth processes. It could be embedded into a game theoretic framework to model competitive forces explicitly. And it could include more behavioral aspects, such as the 'not-invented-here'-syndrome. Also, more empirical research would be useful, particularly in those industries where research does not have much of a tradition. Could these industries benefit from research, or are there good reasons why research has been neglected? Hopefully, these and other questions will be taken up in the near future.

Appendix

Basic research expenditures as a percentage of total research and development expenditures in major industrialized countries, 1971-1993

Year	Germany		Japan		U.S.	
	Industry	Government + Higher Education	Industry	Government + Higher Education	Industry	Government + Higher Education
1971	5.7	.	9.1	.	3.2	36.8
1973	4.8	.	6.7	.	3.0	34.9
1975	4.6	.	5.2	.	3.0	36.8
1977	3.1	.	4.7	.	3.1	37.6
1979	4.5	.	4.6	.	3.0	38.9
1981	5.7	59.0	5.2	33.8	3.1	40.3
1983	4.9	59.4	5.7	36.2	3.4	40.9
1985	4.2	59.7	5.9	34.9	3.4	39.1
1987	5.0	59.8	6.6	27.7	3.2	41.9
1989	5.9	59.5	6.4	27.3	3.2	41.7
1991	5.7	58.1	6.8	.	.	43.9
1993	4.8	.	6.7	.	.	45.0

. = data not available

Sources: SV Wissenschaftsstatistik GmbH, Forschung und Entwicklung in der Industrie, Essen, various volumes; Bundesministerium für Bildung, Wissenschaft, Forschung und Technologie, Bundesbericht Forschung 1996, Bonn 1996, p. 81; Science and Technology Agency, Japanese Government, White Paper on Science and Technology, 1995, Fifty Years of Post War Science and Technology in Japan, Tokyo 1996, p. 288; National Science Foundation, Research and Development in Industry, Washington, D.C. various volumes; OECD, Basic Science and Technology Statistics, Paris 1991, 1993, 1995; National Science Board, Science & Engineering Indicators, Washington 1996.

Nominal and real industrial research expenditure in three countries (Germany, Japan, U.S.)

(1) Germany: Data are taken from SV-Wissenschaftsstatistik GmbH. The deflator of Laspeyers-type was specially developed to reflect the research and development infrastructure (Brockhoff, 1994).

(2) Japan: Data are taken from Government of Japan (1996), where both time series of data are given. Deflation with an index of producer prices leads to slightly stronger growth over the period 1977 to 1993. As stated above, we suspect that the earlier data could be unreliable.

(3) U.S.: Data are taken from National Science Board (1996, p. 108). A GDP implicit price deflator is used to convert nominal into real data. Data on the character of research and development work are voluntary. As the share of companies which reported these data declined in the mid-1980s, more estimates had to be made which relied on information collected in earlier years. Since 1987 the procedure to estimate missing data was changed. It tried to arrive at more actual and more reliable data (National Science Foundation, 1993, p. 106). This seems to have raised the estimated expenditure level and to have introduced greater variance into the time series.

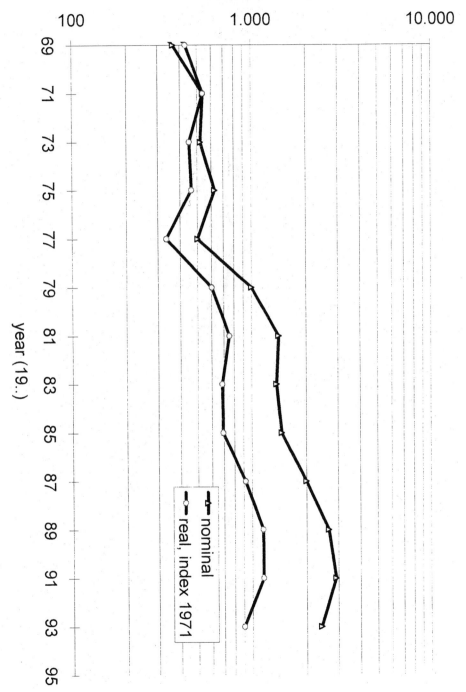

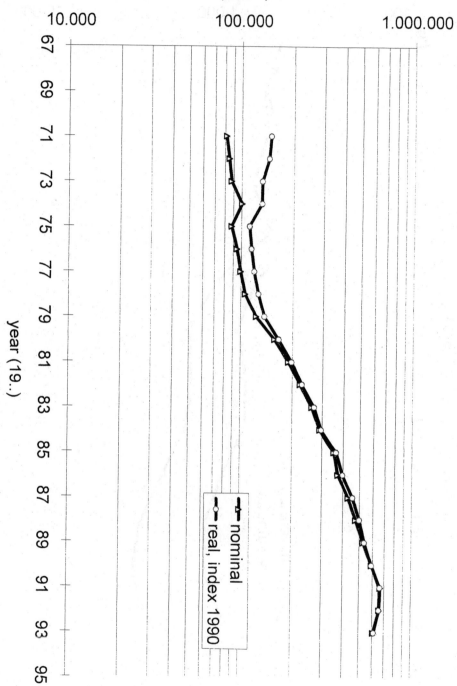

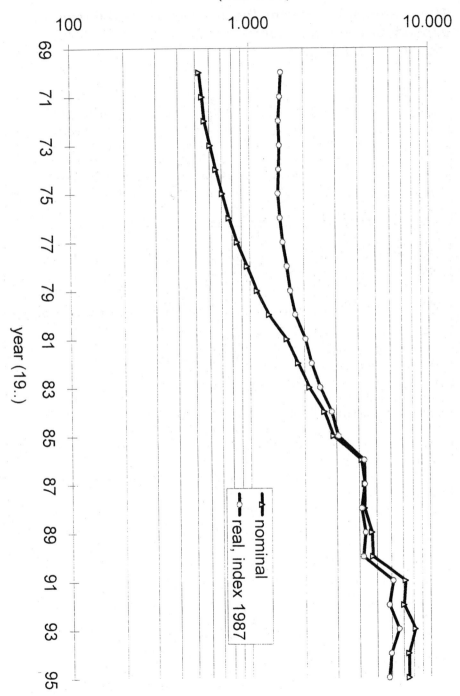

The relationship between the share of industrial research expenditure and the interest rate in Germany for three industries, 1965-1991.

Industry	Chemical/ Plastics/Pharma		Steel construction, Machinery, Automotive		Electrical, Electronics, Optical, Fine mechanics	
Time-lag of the interest rate variable (years)	-1	-2	-1	-2	-1	-2
Time-lag of the research effect (k) (years)	3	6	9	9	4	5
Regression parameter (b)	0.093	0.116	0.085	0.085	0.080	0.087
T-value for the regression parameter	11.8	12.5	8.9	9.1	8.4	8.6
Level of significance for the regression parameter	0.0	0.0	0.0	0.0	0.0	0.0
Coefficient of determination (R-square)	0.927	0.934	0.878	0.881	0.864	0.871

Example of a Mission Statement

Research Division of XYZ-Holding

1. Tasks
Research is an integrated and an integrating part of the entrepreneurial task of our company. It is the leading force in shaping the future of our company. From these statements we derive the following specific tasks:
(a) We want to provide, in time, a reliable technological basis for the key fields in which our companies are engaged. Furthermore, we shall study the cross-sectional technologies v and w.
(b) We want to serve as an information source for the development of new products and business areas.
(c) We want to provide problem solving capacities if new or unconventional questions arise.
(d) By means of proactive thinking, we want to provide guidance for long term development.

2. Management and Leadership
Management and leadership should create a culture that is characterized by goal orientation, creativity, and the commitment to work towards company success. To this purpose:
(a) we want to develop basic knowledge and technological capabilities in all relevant technological areas, which will be transferred to our companies and business units in the form of agreed-upon projects; this should support them in occupying a leading role vis-à-vis their competitors
(b) we want, moreover, to make this knowledge available for use in the strategic planning processes of the business units and the holding company
(c) we pledge to seek intensive information exchange with the business units and at all levels regarding the goals and tasks of the research division
(d) all managers within the research division shall concentrate on fostering creativity and targeting efforts to meet the agreed-upon

objectives; leadership and cooperation shall be object-oriented and not restricted by hierarchies
(e) we will remain open to new scientific ideas, and will try to cooperate with the scientific community to the company's best advantage
(f) we recognize the necessity for a rapid transfer of research results and shall support this transfer
(g) we shall conduct our human resource development such that our employees can meet the challenges of top performance, support the goals of the company, and are creative enough to discover application potentials for new technologies and try to realize these potentials
(h) we shall act according to the credo: Success is not achieved by completing a research project as planned, but by the competitive advances that are enabled by the project!

Literature

Abelson, P.H., Science and Technology Policy, Science, Vol. 267, 27 January 1995, p. 435.

Agassi, J., The Confusion between Science and Technolgy in the Standard Philosophies of Science, Technology and Culture, Vol. VII, 1966, pp. 348-366.

Allen, T. J., Fusfeld, A. K., Design for Communication in the Research and Development Lab, Technology Review, Vol. 78, 1976, 6, pp. 64-71.

Aoki, M., Information, Incentives and Bargaining, Cambridge (Cambridge Univ. Press) 1988.

Armstrong, J., Reinventing Research at IBM, in: Rosenbloom, R.S., Spencer, W.J., Engines of Innovation. U.S. Industrial Research at the End of an Era, op. cit., pp. 151-154,.

Arrow, K., Economic Welfare and the Allocation of Resources for Invention, in: Nelson, R.R., Edt., The Rate and Direction of Inventive Activity, Princeton, NJ (Princeton University Press) 1962, pp. 609-625.

Arthur D. Little, Inc., Basic Research in the Navy, NONR - 2516(00), 1959.

Asakawa, K., External-Internal Linkages and Overseas Autonomy-Control Tension: The Management Dilemma of the Japanese R&D in Europe, IEEE Transactions on Engineering Management, Vol. 43, 1996, pp. 24-32.

Ayres, R., Barriers and Breakthroughs, Technovation, Vol. 7, 1988, pp. 87-115.

Bardenhewer, J., Integrating the Scientific Environment with Industrial research - Results from a European-Japanese Survey and Implications, in: Company of the Future, 1996a.

Bardenhewer, J., Research Elasticities - Empirical Evidence from Delphi-styled Interviews, in: Company of the Future, 1996b.

Bender, A., Simulationsmodell für die FuE-Budgetierung in einem Unternehmen der Chemieindustrie, Manuscript 1997.

Berthold, K., Forschung und Entwicklung in der deutschen Großindustrie. Eine empirische Untersuchung der Forschungs- und Entwicklungstätigkeit der größten Industrieunternehmen in der Bundesrepublik Deutschland unter besonderer Berücksichtigung der Grundlagenforschung, Diss. Mannheim 1968.

Betz, F., Targeted Basic Research. Industry-University Partnerships, in: Gaynor, G.H., Edt., Handbook of Technology Management, New York (McGraw Hill) 1996, Chapter 8.

Binder, V., Kantowsky, J., Technologiepotentiale. Neuausrichtung der Gestaltungsfelder des Strategischen Technologiemanagements. Wiesbaden (Deutscher Universitäts Verlag) 1996.

Bosomworth, C. E., Sage, B. H., jr., How 26 Companies Manage their Central Research, in: Research Technology Management, Vol. 38, 1995, May/June, pp. 32-40.

Bridenbaugh, P.R., The Future of Industrial R&D, or, Postcards from the Edge of the Abyss, in: Rosenbloom, R.S., Spencer, W.J., Edts., Engines of Innovation, U.S. Industrial Research at the End of an Era, op. cit., pp. 155-163.

Brockhoff, K., Forschungsprojekte und Forschungsprogramme. Ihre Bewertung und Auswahl. 2nd ed., Wiesbaden 1973.

Brockhoff, K., Delphi-Prognosen im Computerdialog, Tübingen (Mohr-Siebeck) 1979.

Brockhoff, K., A Simulation Model of R&D Budgeting, R&D Management, Vol. 19, 1989, pp. 265-276.

Brockhoff, K., Stärken und Schwächen industrieller Forschung und Entwicklung. Befragungsergebnisse aus der Bundesrepublik Deutschland. Stuttgart (Poeschel Verlag) 1990.

Brockhoff, K., Forschung und Entwicklung. Planung und Kontrolle, 4th ed., München/Wien (Oldenbourg Verlag) 1994.

Brockhoff, K., Value Generation by Industrial Research, Technovation, Vol. 15, 1995a, pp. 591-600.

Brockhoff, K., Zur Theorie des externen Erwerbs neuen technologischen Wissens, Zeitschrift für Betriebswirtschaft, Supplement 1/1995b, pp. 27-42.

Brockhoff, K., Technology Management in the Company of the Future, Technology Analysis & Strategic Management, Vol. 8, 1996a, pp. 175-189.

Brockhoff, K., Synthesekautschuk: Strategische Aspekte von Stop-and-Go-Entscheidungen in der Entwicklung, in: Brockhoff, K., Management von Innovationen, Wiesbaden (Gabler) 1996b, pp. 125-138.

Brockhoff, K., Schmaul, B., Organization, Autonomy, and Success of Internationally Dispersed R&D Facilities, IEEE Transactions on Engineering Management, Vol. 43, 1996, pp. 33-40.

Bundesministerium für Forschung und Technologie, Edt., Deutscher Delphi-Bericht zur Entwicklung von Wissenschaft und Technik, Bonn 1993.

Bundesministerium für Bildung, Wissenschaft, Forschung und Technologie, Bundesbericht Forschung 1996, Bonn 1996.

Burgelman, R.A., Sayles, L.R., Inside Corporate Innovation, New York 1986.

Chester, A. N., Measurements and Incentives for Central Research, Research-Technolgy Management, Vol. 38, 1995, pp. 14-22.

Chiesa, V., Managing the Internationalization of R&D Activities, IEEE Transactions on Engineering Management, Vol. 43, 1996, pp. 7-23.

Cohen, W.M., Levinthal, D.A., Absorptive Capacity: A New Perspective on Learning and Innovation, Administrative Science Quarterly, Vol. 35, 1990, pp. 128-152.

Collins, P., Wyatt, S., Citations in patents to the basic research literature, Research Policy, Vol. 17, 1988, pp. 65-74.

Coombs, R., Core competencies and the strategic management of R&D, R&D Management, Vol. 26, 1996, pp. 345-370.

de Meyer, A., Mizushima, A., Global R&D Management, R&D Management, Vol. 19, 1989, pp. 135-146.

Dertouzos, M., Lester, R., Solow, R., Edts., Made in America. Regaining the Competitive Edge, Cambridge, Mass. (MIT Press) 1989.

Deutsche Shell AG, Geschäftsbericht 1966.

Domsch, M., Gerpott, H., Gerpott, T.J., Technologische Gatekeeper in der industriellen F&E, Stuttgart (Poeschel) 1989.

Domsch, M., Jochum, E., Peer assessment in industrial R&D departments, R&D Management, Vol. 13, 1983, pp. 143-154.

Dvir, D., Segev, E., Shenhar, A., Technology's Varying Impact on the Success of Strategic Business Units within the Miles and Snow Typology, Strategic Management Journal, Vol. 14, 1993, pp. 155-162.

Echterhoff-Severitt, H., et al., Forschung u. Entwicklung im Wirtschaftssektor in drei Jahrzehnten, Essen (Stifterverband für die Deutsche Wissenschaft) 1988.

Eggers, O., Funktionen und Management der Forschung in Unternehmen, Wiesbaden (Deutscher Universitäts-Verlag) 1997.

EIRMA, Edt., Industry's Needs fo Basic Research, Working Group Report No. 23, Paris (EIRMA) 1982.

Erkner, P., Wachsen im Wettbewerb. Eine Zeitgeschichte der Continental AG (1971-1996) anläßlich des 125-jährigen Firmenjubiläums, Düsseldorf (Econ Verlag) 1996.

Ernst, H., Patenting strategies in the German mechanical engineering industry and their relationship to company performance, Technovation, Vol. 15, 1995, pp. 225-240.

Ernst, H., Patentinformationen für die strategische Planung von Forschung und Entwicklung, Wiesbaden (Deutscher Universitäts Verlag) 1996.

Ernst, H., Industrial Research as a Source of Important Patents, Manuscript, Kiel 1996.

Feibleman, J. K., Pure Science, Applied Science, Technolgy, Engineering: An Attempt at Definitions, Technolgy and Culture, Vol. II, 1961, pp. 305-317.

Foos, C., Teamgeist schlägt Geld, TopBusiness, April 1995, pp. 92-96.

Foray, D., Knowledge Distribution and the Institutional Infrastructure: The Role of Intellectual Property Rights, in: Albach, H., Rosenkranz, S., Intellectual Property Rights and Global Competition, Berlin (Sigma Verlag) 1995, pp. 75-117.

Furnas, C.C., The Philosophy and Objectives of Research in Industry, in: Furnas, C.C., Edt., Research in Industry, 5th ed., Princeton/N.J. (Princeton University Press) 1958, pp. 1-14.

General Accounting Office, National Laboratories: Are Their R&D Activities Related to Commercial Product Development? Washington (GAD) PEMD-95-2, 1995.

Goldman, J. E., McKenzie, L. M., Management of Interface Problems Between Basic an Applied Research, in: Yovits, M. C., et al., Edts., Research Program Effectiveness, New York (Gordon and Breach) 1966, pp. 1-12.

Government of Japan, Science and Technology Basic Plan, July 2, 1996.

Graham, M.B.W., Pruitt, B.H., R&D for Industry. A century of technical innovation at Alcoa. Cambridge (Cambridge University Press) 1990.

Griliches, Z., Productivity, R&D, and Basic Research at the Firm Level in the 1970's, American Economic Review, Vol. 76, 1986, pp. 141-154.

Grossmann, G., Helpman, E., Innovation and Growth in the Global Economy, Cambridge, Mass. (MIT Press) 1991.

Grupp, H., Schmoch, U., Wissenschaftsbindung der Technik, Heidelberg (Physica Verlag) 1992.

Gupta, A.K., Wilemon, D., Improving R&D/Marketing Relations: R&D's Perspective, R&D Management, Vol. 20, 1990, pp. 277-290.

Gwynne, P., 100 Years Small: Managing Innovation at Reilly Industries, Research Technology Management, Vol. 39, 1996, 6 / pp. 39-43.

Hamilton, D.P., Industry Steps in to Fill the Gap in Basic Research, Science, Vol. 258, 1992, 23.10., pp. 570-571.

Hauschildt, J., Innovationsmanagement, München (Vahlen Verlag) 1993.

Hounshell, D.A., Smith, J.K., jr., Science and Corporate Strategy. Research and Development at Du Pont 1908 to 1980, Cambridge (Cambridge University Press) 1989.

Ifo-Institut für Wirtschaftsforschung, Stand und Entwicklungstendenzen der Unternehmensplanung in der Bundesrepublik Deutschland, Vertraulicher Informationsbrief, Februar 1990, Nr. 15.

IIT Research Institute, TRACES, Technology in Retrospect and Critical Events in Science, (National Science Foundation) Vol. 1, 1968; Vol. 2, 1969.

Jensen, M. C., The Modern Industrial Revolution, Exit, and the Failure of Internal Control Systems, Journal of Finance, Vol. 48, 1993, pp. 831-880.

Katz, R., Allen, T.J., Investigating the not invented here syndrome: A look at the performance, tenure, and communication patterns of 50 R&D project groups, R&D Management, Vol. 12, 1982, pp. 7-19.

Katz, R., Tushman, M.L., A Longitudinal Study of the Effects of Boundary Spanning Supervision on Turnover and Promotion in Research and Development, Academy of Management Journal, Vol. 26, 1983, pp. 437-456.

King, R.T., jr., How a Drug Firm Paid for University Study, then Undermined it, Wall Street Journal, 25 April 1996, p.1.

Kinzel, A.B., Basic Research in Industry, National Science Foundation Edt., Proceedings of a Conference on Academic and Industrial Basic Research, Washington/D.C. (National Science Foundation) 1961, pp. 15 et seq.

Kline, S.J., Rosenberg, N., An Overview of Innovation, in: Landau, R., Rosenberg, N. Edts., The Positive Sum Strategy, Washington, D.C. 1986, pp. 275-305.

Kolatek, C., Das Management von Forschungs- und Entwicklungsaktivitäten in japanischen Unernehmen, in: Albach, H., Edt., Innovationsmanagement. Theorie und Praxis im Kulturvergleich. Wiesbaden (Gabler) 1990, pp. 177-213.

Kümmerle, W., Praxisorientierte Grundlagenforschung im Dienste japanischer Wettbewerbsstrategie, Handelsblatt, July 8, 1993, p. 24.

Lange, V., Technologische Konkurrenzanalyse, Wiesbaden (Deutscher Universitäts Verlag) 1994.

Leonard-Barton, D., Wellsprings of Knowledge: Building and Sustaining the Sources of Innovation, Boston (Harvard Business School Press) 1995.

Leonard-Barton, D., Pisano, G., Monsanto's March into Biotechnology, Harvard Business School Case, 9-960-009, 1993.

Link, A.N., Basic Research and Productivity Increase in Manufacturing: Additional Evidence, American Economic Review, Vol. 71, 1981, pp. 1111-1112.

Linstone, H. A., Turoff, M., Edts., The Delphi Method. Techniques and Applications. Reading, Mass. (Addison-Wesley Publishing Co.) 1975.

MacCrimmon, K.R., Wehrung, D.A., Taking Risks. The Management of Uncertainty. New York/London (Free Press) 1986.

Mansfield, E., Basic Research and Productivity Increase in Manufacturing, American Economic Review, Vol. 70, 1980, pp. 863-873.

Mansfield, E., Academic research and industrial innovation, Research Policy, Vol. 20, 1991, pp. 1-12.

Mansfield, E., Lee, J. Y., The modern university: contributor to industrial innovation and recipient of R& D support, Research Policy, Vol. 25, 1996, pp. 1047-1058.

Marshall, A., Industry and Trade, 2nd ed., London (MacMillan & Co.) 1927.

Martin, B. R., Irvine, J., Research Foresight. Priority Setting in Science. London, New York (Pinter Publ.) 1989.

Martin, B.R., Irvine, J., Assessing basic research. Some partial indicators of scientific progress in radioastronomy, Research policy, Vol. 12, 1983, pp. 61-90.

Medcof, J.W., A Taxonomy of Internationally Dispersed Technology Units and its Application to Management Issues, Working Paper, McMaster Univ., Hamilton, Canada, 1996.

Mehrwald, H., The 'Not Invented Here'(NIH)-Syndrome in the Inter-Organisational Technolgy Transfer Process, Working Paper, presented at the 6th European Doctoral Summer School in Technology Management Manchester, U.K., 19.-30.Aug., 1996.

Meyer-Krahmer, F., The German R&D system in transition: Empirical results and prospects of future development, Research Policy, Vol. 21, 1992, pp. 423-436.

Meyers, L. A., Information Systems in Research and Development: The Technological Gatekeeper Reconsidered, R&D Management, Vol. 13, 1983, pp. 199-206.

Mittelstraß, J., Wissenschaftstheoretische Bemerkungen zum Forschungsbegriff und zur Forschungsorganisation. In: Stifterverband für die deutsche Wissenschaft, edt., Von der Hypothese zum Produkt, Essen (Stifterverband für die Deutsche Wissenschaft) 1995, pp. 18-24.

Moore, G. E., Some Personal Perspectives on Research in the Semiconductor Industry, in: Rosenbloom, R.S., Spencer, W.J., Edts., Engines of Innovation. U.S. Industrial Research at the End of an Era, op. cit., pp. 165-174.

Morone, J., Technology and Competitive Advantage - The Role of General Management, Research-Technology Management, Vol. 36, 1993, pp. 16-25

Nanaka, I., Takeuchi, H., The Knowledge-Creating Company. How Japanese Companies Create the Dynamics of Innovation. New York, Oxford (Oxford Univ. Press) 1995.

Narin, F., Carpenter, M. P., Woolf, P., Technological Performance Assessments Based on Patents and Patent Citations, IEEE Transactions on Engineering Management, Vol. EM 31, 1984, pp. 172-183.

Narin, F., Noma, E., Perry, R., Patents and indicators of corporate technological strength, Research Policy, Vol. 16, 1987, pp. 143-155.

Nash, J. C., Nonlinear Parameter Estimation, New York (Marcel Dekker) 1987.

National Science Board, Science & Engineering Indicators - 1996, Washington (U.S. Government Printing Office) 1996.

National Science Foundation, Methodology of Statistics on Research and Development, Washington (NSF) 1959.

National Science Foundation, Research and Development in Industry, Washington, D.C. (NSF 96-304) 1993.

Nelson, R., The Simple Economics of Basic Scientific Research, Journal of Political Economy, Vol. 67, 1959, pp. 297-306.

Nelson, R., Understanding Technical Change as an Evolutionary Process, Amsterdam (North Holland) 1987.

Neukirchen, H., Forschung als professionelle Dienstleistung der Holding, Welt am Sonntag, 38, September 22, 1996, p. 59.

OECD, Frascati Manual 1992, Proposed Standard Practice for Surveys of Research and Experimental Development, Paris (OECD), 1992.

Ordover, J., A Patent System for both Diffusion and Exclusion, Journal of Economic Perspectives, Vol. 5, 1991, pp. 43-60.

Pavitt, K., What makes basic research economically useful? Research Policy, Vol. 20, 1991, pp. 109-120.

Pearce, R., Papanastassiou, M., R&D networks and innovation: decentralized product development in multinational enterprises, R&D Management, Vol. 26, 1996, pp. 315-334.

Pearson, A., Brockhoff, K., von Boehmer, A., Decision parameters in global R&D management, R&D Management, Vol. 23, 1993, pp. 249-262.

Perlitz, M., Hollax, M., Schrank, R., Schug, K., Accountability of Technology, in: Company of the future, 1996.

Peyer-Roche, H. C., Geschichte eines Unternehmens 1896 - 1996, Basel (Roche) 1996.

Pieper, U., Vitt, J., Die Messung der technologischen Verwandtschaft von Akquisitionsunternehmen, Manuskript, Kiel 1996.

Pisano, G. P., The R&D Boundaries of the Firm: An Empirical Analysis, in: Administrative Science Quaterly, Vol. 35, 1990, pp. 153-176.

Prahalad, C. K., The role of core competencies in the corporation, Research-Technology Management, Vol. 36, 1993, 6/pp. 40-47.

Prahalad, C. K., Hamel, G., Core Competence of the Corporation, Harvard Business Review, Vol.68, 1990, pp. 79-91.

President's Commission on Industrial Competitiveness. Global Competition. The New Reality. Vol. I, II, Washington/D.C. 1985.

Price, D. J. Desolla, Little Science, Big Science, New York 1963.

Robb, W. L., How good is our Research? Research Technology Management, March/April 1991, pp. 16-21.

Ronstadt, R. C., International R&D: The establishment and evolution of research and development abroad by seven US multinationals, Journal of International Business Studies, Vol. 9, 1978, pp. 7-24.

Rosenberg, N., Why do firms do basic research (with their own money)? Research Policy, Vol. 19, 1990, pp. 165-174.

Rosenbloom, R.S., Kantrow, A.M., The nurturing of corporate research, Harvard Business Review, Vol. 60, January/February 1982, pp. 115-123.

Rosenbloom, R. S., Spencer, W. J., Engines of Innovation. U.S. Industrial Research at the End of an Era, Cambridge, Mass. (Harvard Business School Press) 1996.

Rubenstein, A. H., Liaison Relations in Research and Development, IRE Transactions on Engineering Management, Vol. EM-4, 1957, June, pp. 72-78.

Rubenstein, A. H., Barth, R. T., Douds, Ch. F., Ways to Improve Communications Between R&D Groups, Research Management, 1971, November, pp. 49-59.

Rubner, J., Forschen auf der Insel der Seligen, Süddeutsche Zeitung, September 13, 1996.

Ruedi, A., Lawrence, P., Aerospace Systems, Harvard Business School case 9-474-164, as reproduced in: Burgelman, R.A., Maidique, M.A., Wheelwright, S.C., Strategic Management of Technology and Innovation, 2nd ed., Chicago et al. (Irwin) 1995, pp. 507-521.

Schewe, G., Imitationsmanagement, Stuttgart (Poeschel Verlag) 1992.

Schmitt, J., Bell Labs research may suffer, USA Today, September 22, 1995.

Schrader, S., Sattler, H., Zwischenbetriebliche Kooperation: Informaler Informationsaustausch in den USA und Deutschland, Die Betriebswirtschaft, Vol. 53, 1993, pp. 587-606.

Science and Technology Agency, Government of Japan, White Paper on Science and Technology 1995.

Smith, A., The Wealth of Nations, London (J.M.Dent & Sons) 1910.

Speiser, A., Forschung in Hochschule und Industry: Wechselwirkungen zwischen Wissenschaft und Technik, in: Bundesministerium für Wissenschaft und Forschung, Edt., Forschungstheorie Forschungspraxis, Wien/New York (Springer Verlag) 1971, pp. 7-20.

Stand und Entwicklungstendenzen der Unternehmensplanung in der Bundesrepublik Deutschland, Ifo-Vertraulicher Informationsbrief, Februar 1990, Nr. 15, pp. 1-5.

Steinmueller, W. E., Basic Research and Industrial Innovation, in: Dodgson, M., Rothwell, R., The Handbook of Industrial Innovation, London (E.Elgar) 1994, pp. 54-66.

SV Wissenschaftsstatistik GmbH, Forschung und Entwicklung in der Industrie, Essen, various volumes.

Takeda, Y., Management of the Company of the Future - Case study of a global high-tech company 'Company X', Company of the Future, Core Member Meeting, Tokyo, Feb. 25, 1996.

Tushman, M. L., Scanlan, D. A., Characteristics and External Orientation of Boundary-Spanning Individuals, Academy of Management Journal, Vol. 24, 1981, pp. 83-98.

U.S. General Accounting Office, The Federal Role in Fostering University-Industry Cooperation, Report GAO/PAD-83-22, May 25, 1983, p.24.

Uttal, B., The Lab that ran away from Xerox, Fortune, Sept. 5, 1983.

Van Alstyne, M., Brynjolfsson, E., Could the Internet Balkanize Science? Science, Vol. 274, 1996, 29 November, pp. 1479-1480.

Van Vianen, B. G., Moed, H. F., Van Raan A. F. J., An exploration of the science base of recent technology, Research Policy Vol. 19, 1990, pp. 61-81.

von Boehmer, A., Internationalisierung industrieller Forschung und Entwicklung, Wiesbaden (Deutscher Universitäts Verlag) 1995.

Wagstyl, S., An open market in industrial research, Financial Times, Oct. 22, 1996.

Warschkow, K., Organisation und Budgetierung zentraler FuE-Bereiche, Stuttgart (Poeschel Verlag) 1993.

Weitzel, G.U., Unternehmensdynamik und globaler Innovationswettbewerb, Wiesbaden (Gabler Verlag) 1996.

Weule, H., et al., Zusammenarbeit GFE/Industrie, Stuttgart, Mai 1994.

Yasai-Ardekani, M., Nystrom, P.C., Designs for Environmental Scanning Systems: Tests of a Contingency Theory, Management Science, Vol. 42, 1996, pp. 187-204.

ZVEI (Zentralverband Elektrotechnik und Elektroindustrie e.V.), Technologien im 21. Jahrhundert. Aktionspapier zur Innovationsförderung, Frankfurt 1994a.

ZVEI, Bewertung der Industrierelevanz staatlich geförderter Forschungseinrichtungen im Bereich der Informationstechnik, Frankfurt 1994b.

Other Material:

Business Week, March 30, 1987.

Dresdner Bank, Historische Statistische Reihen, Durchschnittsrendite festverzinslicher Wertpapiere, May 1991.

du., Hoechst verlagert immer mehr Forschung in das Ausland, Frankfurter Allgemeine Zeitung, September 14, 1996.

hra, Erfolgreiche Produktion eines blau strahlenden Halbleiterlasers, Frankfurter Allgemeine, No. 29, February 3, 1996, p. 19.

Science, Oct. 25, 1992, p. 571

Shell schließt ihre Laboratorien in Birlinghoven am 31. Oktober, Siegkreis-Rundschau, August 9, 1966.

Siemens Forschung, 47000 Mitarbeiter in Laboratorien tätig, Handelsblatt, January 17, 1992.

Druck: Strauss Offsetdruck, Mörlenbach
Verarbeitung: Schäffer, Grünstadt